Cutting Edge Technologies

National Academy
of
Engineering

NATIONAL ACADEMY PRESS
Washington, D.C. 1984

National Academy Press • 2101 Constitution Avenue, NW • Washington, DC 20418

The National Academy of Engineering is a private organization established in 1964. It shares in the responsibility given the National Academy of Sciences under a congressional charter granted in 1863 to advise the federal government on questions of science and technology. This collaboration is implemented through the National Research Council. The National Academy of Engineering recognizes distinguished engineers, sponsors engineering programs aimed at meeting national needs, and encourages education and research.

Library of Congress Cataloging in Publication Data

Main entry under title:

Cutting edge technologies

Papers of the Symposium on Cutting Edge Technologies convened in 1983 by the National Academy of Engineering.
1. High technology—Congresses. I. National Academy of Engineering. II. Symposium on Cutting Edge Technologies (1983: Washington, D.C.)
T6.C88 1984 620 84-19007

ISBN 0-309-03489-2

Printed in the United States of America

Symposium Steering Committee

Chairman

N. BRUCE HANNAY, Foreign Secretary, National Academy of Engineering; Vice-President, Research and Patents, Bell Laboratories (retired)

Members

RALPH E. GOMORY, Vice-President and Director of Research, IBM Corporation

RALPH HARDY, Director, Life Sciences Division, E. I. du Pont de Nemours & Co., Inc.

MILTON PIKARSKY, Director, Transportation Research, IIT Research Institute

ALBERT R. C. WESTWOOD, Corporate Director, Research and Development, Martin Marietta Corporation

Contents

Preface

Rapid change is occurring in many scientific, engineering, and technological fields. As the rate of change accelerates, specialists in particular fields find it more difficult to keep abreast of the increase in knowledge and technique. It is even more difficult for scientists and engineers to understand how developments outside their fields of specialty are likely to affect what they do and how they do their work. The symposium on Cutting Edge Technologies, convened by the National Academy of Engineering in connection with its Annual Meeting in November 1983, was organized to provide members of the Academy and a wider audience interested in the present state of development of various cutting edge technologies with an overview of the general trends taking place in a number of fields.

In a symposium such as this, it was impossible to address the problems of technological development in a comprehensive way, and hence a limited number of technologies were identified for discussion, recognizing that there are many others that could just as easily have been candidate topics for a symposium. The topics selected range from those that have been consistently in the forefront of national attention to those that have had less exposure but are nonetheless central to the national welfare and industrial competitiveness of the United States. The symposium topics that were selected include those dealing with the computers of the future and new frontiers in biotechnology—two areas that have been the center of much media as well as professional attention—and those on advances in structural materials and transportation technology—two vital areas that have received less public attention.

We hope that the papers in this volume can place in perspective the present state of development of individual technologies and how they are likely to evolve over the next several decades. If the individual and collective effect of the papers also communicates the vitality of the multiple streams of technological development taking place, then a major goal of the symposium will have been achieved. Just as importantly, however, the collection of papers in this volume foreshadows the changing technological environment that will embrace our society in the years ahead.

ROBERT M. WHITE
President
National Academy of Engineering

Part I

Computers of the Future

Introduction

RALPH E. GOMORY

In the first of three papers on computers of the future, James D. Meindl discusses, or rather gives the reader a feeling for, the hardware. Because there are too many elements of hardware to permit covering each specifically, he discusses what is probably the key component, the field effect transistor (FET).

In particular, Dr. Meindl discusses where FETs in ultra large scale integration are going. To do that, he considers a hierarchy of theoretical and practical limits ranging from those imposed by the laws of physics to those imposed by properties of materials, devices, circuits, and systems. To project future developments, he extrapolates from past decreases in the feature size of FETs, increases in the die or chip size, and improvements in the "cleverness" of circuit design. He also presents some thought-provoking analogies between the development of iron technology during the industrial revolution for our structural needs and the ongoing development of silicon technology for our electronic needs.

In the second paper Herbert Schorr explains how hardware is put together into systems of orthodox design. "Systems" will include the hardware and the software that makes it all work together. The growth will come from both the continued improvements in technology that will reduce cycle times and the improvements in the design of processors that will reduce the number of cycles required per instruction.

As Dr. Schorr indicates, we will continue to see multiple-machine environments combining uniprocessors of ever-increasing power. Microprocessors and powerful workstations offer the possibility of alternative architectures to today's configurations where most processing is done by the central host. System interconnection and distributed processing will play an increasingly important role.

And finally, Michael L. Dertouzos has the challenging task of going beyond the orthodox machines and the orthodox architectures; he explains how one might put together machines of a totally different design. Computer networks provide means for autonomous, geographically distributed computer systems, while tightly coupled multiprocessor systems provide for enormous, affordable, processing power.

He also discusses what has been perhaps the great bottleneck of the field, which is the development of the software—not its running, not its structure, but how to bring it into existence. He envisions the transition of the software development process from the artisan to the mass production era. Programs will be more intelligent and flexible for applications.

So, in brief, this set of papers discusses the hardware, the area in which rapid advances made computers economically important and make them more important every year; the orthodox systems; and the unorthodox systems and the creation of software.

Ultra Large Scale Integration and Beyond

JAMES D. MEINDL

The objective of this paper is to project the future of integrated electronics by studying the theoretical, practical, and analogical limits that govern its progress. The limits governing ultra large scale integration (ULSI) of 10 million to 1 billion transistors in a single chip of silicon (Si) can be organized in a hierarchical matrix, as illustrated in Figure 1.

At the first level of the hierarchy are fundamental limits, which are immutable laws of nature; they cannot be changed. At the second level are material limits, which are specific to composition but do not change frequently in practice. Silicon has been the keystone material of integrated electronics for the past two decades, and this is unlikely to change during the next two. At the third level, device limits depend upon both the material properties and configuration of ULSI components. Consequently, these limits are useful in projecting the smallest possible dimensions of structures such as insulated gate field effect transistors (IGFETs). At the fourth level are circuit limits—unique because they retain both a complete physical description and a definition of the information-processing function of a group of components and interconnections. Moreover, circuit limits describe device performance in a realistic operating environment, not in sterile isolation. Consequently, the circuit level of the hierarchy is the most appropriate one for projecting the smallest allowable dimensions of ULSI structures for specific purposes. The fifth level, the system limits, can be elaborated into several discrete steps reflecting the logic design, architecture, instruction set,

	THEORETICAL	PRACTICAL	ANALOGICAL
5. SYSTEM			
4. CIRCUIT			
3. DEVICE			
2. MATERIAL			
1. FUNDAMENTAL			

FIGURE 1 Hierarchical matrix of limits governing integrated electronics.

algorithms, and application of a particular ULSI configuration. System limits are the most numerous and nebulous set of the hierarchy. However, because opportunities for integration at each of the five hierarchical levels are constrained by the limits of all preceding levels, system limits represent the most profoundly important set.

As will be discussed in the next two major sections of this paper, both theoretical and practical limits are included at each level of the hierarchy. For example, at the fundamental level thermal fluctuations impose a theoretical limit on switching energy of several kT (where k is Boltzmann's constant and T is absolute temperature) that is further restricted by practical constraints on cooling temperature. With regard to the second, or materials, level, although theory suggests use of high-mobility GaAs or InP materials, for overwhelming practical reasons Si dominates integrated electronics. At the third level, avoidance of drain-to-source junction punch-through determines a theoretical minimum channel length for IGFET devices that may never become practical because of manufacturing limitations of microlithographic technology. At the fourth level, practical supply voltage standards may prevent reaching theoretical circuit limits on the power-delay product of complementary metal-oxide-semiconductor (CMOS) technology. And, finally, common clock skew illustrates a simple system limit depending on interconnect time delay.

The totality of practical limits is described by three parameters that collectively measure the overall rate of progress of integrated electronics: (1) minimum feature size, (2) die area, and (3) packing efficiency of a complex chip. Packing efficiency has been described by Moore[1] as "cleverness." Its precise definition is the number of transistors per minimum feature area. The combined time derivatives of these parameters determine the rate of change of the total number of components per chip, the central measure of progress in integrated electronics.

As discussed in the section on analogical limits, an intriguing projection of the long-term future of integrated electronics can be developed by comparison of the mature industrial revolution with the modern information revolution. A correspondence between structural materials in the industrial revolution and electronic materials in the information revolution suggests a set of long-term analogical limits on ULSI. The observation that technological advances in different fields often have followed similar patterns of development provides a basis for introducing these limits.

THEORETICAL LIMITS

This section projects the smallest allowable dimensions for integrated structures using as relevant criteria circuit limits associated with IGFETs, polycrystalline silicon or polysilicon resistors, and interconnections.

Transistors

Fundamental, material, and device limits in digital electronics have been surveyed by Keyes.[2] Dennard et al.[3] described a constant electric field scaling theory for IGFETs. A later publication[4] discussed constant voltage and quasi-constant voltage scaling. Most recently, experimental IGFETs with effective channel lengths of approximately 0.15 microns were reported by Fichtner et al.[5]

Constant electric field (CE) scaling of IGFETs begins with the definition of a device scaling factor $S > 1$. All lateral and vertical device dimensions are scaled down by the same factor $1/S$. In addition, drain supply voltage is scaled as $1/S$ in order to maintain a constant electric field intensity and consequently undiminished safety margins for device operation. The principal benefits of CE scaling are a device delay time that decreases as $1/S$, a power density that remains constant, a packing density that increases as S^2, and a power-delay product that decreases as $1/S^3$.

In comparison, constant voltage (CV) scaling also calls for reducing device dimensions by the factor $1/S$, but supply voltage remains constant in order to maintain compatibility with established standards. The principal benefits of constant voltage scaling are device delay time, packing density, and power-delay product that scale as $1/S^2$, S^2, and $1/S$, respectively. A disadvantage of CV scaling is a power density that increases as S^3 and therefore aggravates the already serious heat-removal problem in integrated electronics. In practice, a compromise between CE and

CV scaling (roughly equivalent to quasi-constant voltage [QCV] scaling) has prevailed and can be expected to continue. This compromise includes scaling different dimensions at different rates.

At this time the state-of-the-art minimum feature size of commercial very large scale integration (VLSI) is approximately 2.0 microns. It is pertinent to ask, "Can we project future ULSI structures with 0.2 micron minimum feature sizes?"[6] To respond on the basis of circuit scaling limits, an accurate analytical circuit model for short-channel IGFETs offers marked advantages in physical insight and computational efficiency in comparison with empirical or numerical models. An early IGFET circuit model[7] described long-channel superthreshold device behavior. It was extended to describe subthreshold performance[8] and the coarse effects of short channels via geometric charge sharing approximations.[9] Recent models based upon analytical solution of the two-dimensional (2D) Poisson equation[10,11] effectively account for short-channel and source/drain bias effects. Although these 2D models lack the high accuracy of CAD models of current state-of-the-art IGFETs,[12] their compact representation of 2D effects provides a promising basis for investigation of integrated circuit limits. The 2D models have been employed in studies of enhancement/depletion N-channel MOS transistor (E/D NMOS) and N-channel and P-channel or complementary MOS transistor (CMOS)[11] digital circuits. For conservative design margins, typical results suggest that IGFET channel lengths can be reduced to approximately 0.40 microns in E/D NMOS logic gates, 0.30 microns in CMOS transmission gates, and 0.20 microns in CMOS logic gates. Smaller channel lengths can be projected for more aggressive designs. The dominant mechanism imposing these limits is subthreshold drain current or leakage current, which degrades integrated circuit performance and thereby forces an end to IGFET scaling.

Resistors

Following the pattern of CE IGFET scaling, resistor scaling calls for reducing all device dimensions by the factor 1/S. Since both voltage and current scale as 1/S, resistance remains constant. Consequently, resistivity must decrease as 1/S for CE scaling. (For CV scaling, resistance decreases as 1/S and resistivity as $1/S^2$.)

Because of its advantages in static random-access memory chips, or SRAMs, polysilicon is now the most widely used resistor material in VLSI. Scaling polysilicon resistivity is enhanced by recent advances in modeling carrier transport in this material.[13-15] Salient features of advanced models include identical cubic grain sizes, uniform ionized impurity distribution without segregation at grain boundaries, monoenergetic interface traps, symmetrical semiconductor-to-semiconductor abrupt junctions at grain interfaces, thermionic emission over barriers for $T \geq 27°C$, one-dimensional carrier transport, equally distributed voltage increments across grains, and partially or totally depleted grains. Considering both bulk crystallite and grain boundary contributions to resistivity yields a sinh current-voltage characteristic whose argument varies inversely with the number of grains in the resistor. Hence, linearity is enhanced as the number of grains increases or as the voltage increment per grain decreases. In addition, measurements indicate that the sensitivity of resistivity to fluctuations in doping concentration is inversely related to crystallite size.

The implications of the two preceding properties in terms of resistor scaling are unfavorable. Scaling length by 1/S requires reducing voltage by 1/S in order to maintain the same degree of linearity and reducing the number of grains by 1/S to prevent sensitivity degradation. Thus, QCV scaling tends to degrade linearity, and the small numbers of grains encountered at submicron lengths imply large variances in distributions of resistance values. The unfavorable consequences of scaling polysilicon resistors to lengths smaller than 1.0 micron suggest their replacement with silicon-on-insulator (SOI) transistors fabricated in polysilicon[16] or recrystallized silicon.[17]

Interconnections

Scaling of interconnections entails the definition of both a device scaling factor S and a chip scaling factor $S_c > 1$, since die area tends to increase with time.[18] Again following the pattern of CE scaling, local interconnections (e.g., interconnections within a logic gate or between adjacent gates) are scaled in all dimensions by 1/S. However, long-distance interconnections (e.g., interconnections extending from corner to corner of a die) are scaled by S_c in length and $1/S^2$ in cross-sectional area. Among the results of local interconnect scaling (in combination with CE IGFET scaling) are that the interconnect time constant or response time and voltage drop remain constant and current density increases with S. Since IGFET time delay decreases as 1/S, a constant local interconnect response time assumes increasing importance. Moreover, the salient result of long-distance interconnect scaling is a response time that increases rapidly as $(SS_c)^2$.[19] Long-distance interconnections fabricated with polysilicon or silicide materials now exhibit response times well in excess of individual logic gates. If current trends persist, even aluminum interconnects will surpass logic gates in response time during the mid-1980s. To push back this barrier, greater use of multilevel metal interconnections can reduce average interconnect length.[20] In addition, cascade line drivers and repeaters are useful circuit techniques. A key design criterion is that total interconnect resistance should be less than about 2.3 times the driver transistor resistance.[21] New chip architectures that reduce average interconnect length represent a promising approach at the system level.

Although the preceding discussion of interconnect scaling applies specifically for line-to-substrate parasitic capacitance in combination with CE IGFET scaling, the salient results remain largely unaltered as one extends the analysis to include line-to-line parasitic capacitance, CV IGFET scaling, and scaling different dimensions (e.g., interconnect thickness and field insulator thickness) at different rates.[21]

Overview

A representative overview of the hierarchy of limits governing integrated electronics is obtained by a power-delay plot, as illustrated in Figure 2. The fundamental limits are due to thermal noise and the quantum mechanical uncertainty principle. The material limit on carrier

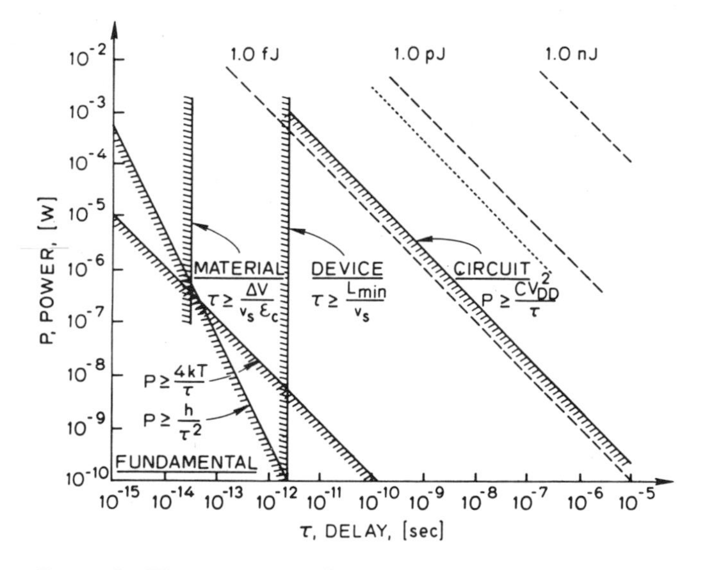

FIGURE 2 Hierarchy of limits illustrated in power-time delay plane including fundamental limits from thermodynamics ($4kT/\tau$) and quantum mechanics (h/τ^2) as well as material, device, and circuit limits. The dotted line represents 1982 state-of-the-art circuitry.

transit time in Si is imposed by scattering limited carrier velocity and critical field strength. The IGFET device limit is set by velocity saturation and punch-through. The circuit limit applies to a CMOS logic gate with minimum-geometry IGFETs. The dotted line in Figure 2 represents the state-of-the-art power-delay product circa 1982.

PRACTICAL LIMITS

Theoretical limits are based on the principles of solid-state science, whereas practical limits depend on manufacturing processes and equipment. The status of the five levels of practical limits that constrain integrated electronics can be summarized in terms of three parameters: minimum feature size, die area, and packing efficiency. Although theoretical studies have established limits on minimum feature size, similar limits on die area and packing efficiency have not been defined.

Feature Size

During the period 1959 to 1975 minimum feature size decreased at a rate of approximately 11 percent per year, resulting in a halving every six years.[1] (Minimum feature size is defined as one-half the minimum pitch, or the average of minimum line width and spacing.) A 50 percent reduction in feature size over six years corresponds to a 4-times increase in components per unit area. In turn, this equates to approximately an 8.7-times increase in components per unit area per decade. Since 1975, minimum feature size has continued to shrink at its previous rate, reaching a value of approximately 2.0 microns in 1982 in state-of-the-art commercial integrated electronics. It has been estimated that optical lithography, the primary tool for wafer exposure, will reach the limits of its capacity for small feature sizes in the range of 0.75 to 0.50 microns.[22,23] Assuming that feature size continues to decrease at its historical rate, optical lithography as practiced in widespread high-volume production should reach its limits in the 1990-1994 period. In this range of time a breakpoint in the feature size versus time curve followed by a reduction in slope may be anticipated. Thereafter, more advanced microlithographic techniques such as E-beams and X-rays should extend IGFET minimum feature size to its theoretical limits in the 0.40- to 0.20-micron range (Figure 3).

Die Size

From 1959 to 1975 integrated circuit die area increased at a rate of approximately 9 percent per year, resulting in a quadrupling of die area and a doubling of the associated die edge every eight years.[1] Since 1975, die area has continued to grow at approximately its previous rate (Figure 4), reaching a "representative range" of about 36 to 64 square millimeters (mm^2) in 1982, depending on the type of circuit, including programmable read-only memories (PROMs), electronically programmable read-only memories (EPROMs), static random-access memories (SRAMs), dynamic random-access memories (DRAMs), microcomputers, image sensors, and gate arrays, within the estimate.

A 100 percent increase in die edge over eight years corresponds to a 4-times increase in components per chip and therefore approximately a 5.7-times increase in components per chip per decade. From 1959 to 1982 the combined effects of decreasing minimum feature size and increasing die size produced an increase of approximately 49 times per decade in the number of components per chip.

In projecting future increases in die size, it is assumed that the historical rate established through optical lithography will continue into the early 1990s. Thereafter, the combined effects of decreasing minimum feature size and increasing die size (i.e., the combined effects of further advances in lithography) might be expected to produce an increase of about 5 times per decade in the number of components per chip. Several arguments can be suggested for this projected 10-times falloff in litho-

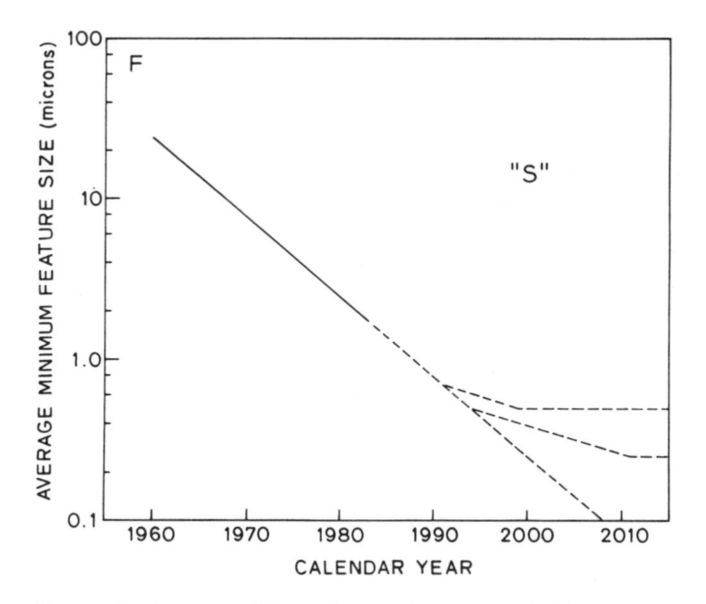

FIGURE 3 Average minimum feature size versus calendar year.

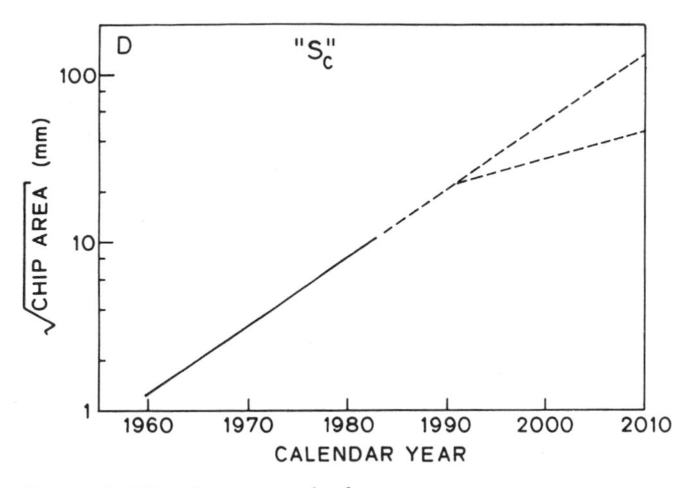

FIGURE 4 Die edge versus calendar year.

graphic productivity increases. The proximity of the theoretical limits on minimum feature sizes will slow advances in ULSI technology. New and more demanding lithographic techniques such as E-beams and X-rays will be necessary to surpass the limits of optics. More severe packaging problems, including interconnections and heat removal, will inhibit die size. Finally, experience already has shown a drastic reduction in the rate of increase of packing efficiency, thus suggesting the possibility of a similar rate reduction due to lithography.

Recent advances in the use of redundancy in large memory chips offer promise of developments in wafer scale integration (WSI) that will depart markedly from historic trends in die size. Although the future course of these efforts has not been projected, one may speculate on a step increase of more than 10 times in monolithic silicon substrate area, compared with the area of a single die, for future integrated systems incorporating wafer scale integration.

Packing Efficiency

The contribution of process, device, and circuit innovations to increasing the number of components per chip can be described by a packing efficiency parameter. To compute a representative value of this contribution, for example, suppose that a 256K DRAM cell occupies an area of 71 square microns with a minimum feature size of 1.5 microns. Moreover, suppose a preceding 65K DRAM cell occupies an area of 256 square microns with a minimum feature size of 2.0 microns.

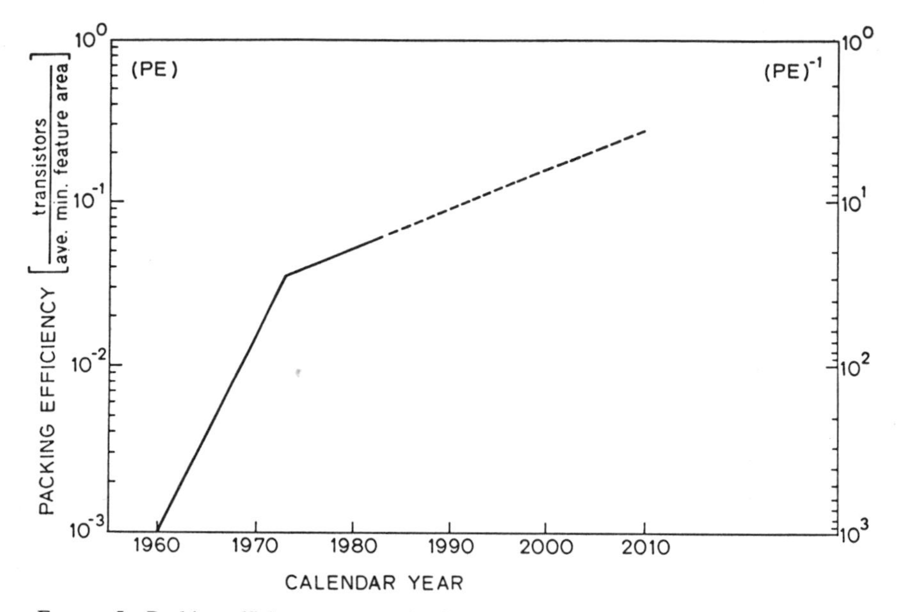

FIGURE 5 Packing efficiency versus calendar year.

The resulting packing efficiency contribution can be computed as $(256/71)(1.5/2.0)^2 = 2.03$ times.

An alternative approach is to observe, for example, that the overall rate of increase of components per chip is 102 times per decade and that the combined contributions of decreasing feature size and increasing die size account for 49 times per decade, thus giving 2.1 times per decade as the packing efficiency contribution. A plot of packing efficiency versus calendar year indicates that about an order-of-magnitude reduction in its rate of increase occurred in the early 1970s (Figure 5).

Number of Components per Chip

From 1959 to 1972/1973 the total number of components per chip increased from 1 to approximately 11,000. This corresponds approximately to an increase of 100 percent annually, or 1,024 times per decade. Given that smaller minimum feature sizes and larger die sizes accounted for 8.7- and 5.7-times-per-decade increases, respectively, packing efficiency improvements must then have produced about a 21-times-per-decade increase in the number of components per chip.

From 1972/1973 to 1981 the number of components per chip increased from approximately 11,000 to 600,000 (e.g., the 256K DRAM) corresponding to an increase of 59 percent per year, 4 times per three years, or 102 times per decade. Since smaller feature sizes and larger die sizes continued to account for 8.7- and 5.7-times-per-decade increases, respectively, during this period packing efficiency improvements produced only a 2.1-times-per-decade increase in the number of components per chip. This drastic reduction of 10 times in the rate of packing efficiency improvements apparently is the result of exhausting relatively easily obtained gains in layout density that marked the first decade or so of integrated electronics.

The packing efficiency contribution of about 2.1 times per decade achieved since 1972/1973 may be maintained through the 1980s and perhaps the 1990s as well. Advances in multilevel (sometimes referred to as three-dimensional, or 3D) integrated circuits will contribute to this advance. However, the problems of wafer scale integration appear to be more readily amenable to solution, e.g., through the use of redundancy, than do those of multilevel circuits with more than two levels of active devices. Thus, assuming that the problems of product definition and design as well as of heat removal[24] are solved, wafer scale integration may become the source of a new factor increasing the number of components per monolithic Si substrate.

Overview

An overview of the preceding discussion of practical limits is illustrated in Figure 6. The projected lower boundary corresponding to line segment E assumes a 0.75-micron limit for optical lithography and a 0.50-micron limit for IGFET minimum feature size. The upper boundary corresponding to D assumes a 0.50-micron limit on optical lithography and a 0.25-micron limit for IGFETs. The expected range of performance lies within these boundaries.

ANALOGICAL LIMITS

Historical studies indicate that technological advances in different fields often follow similar patterns of development. For example, it has been observed that a growth curve that follows an S shape (a slow start, then a rapid rise, followed by a leveling off and obsolescence) is a common pattern.[25,26] This has led to the use of historical analogy as a forecasting tool. An analogy between structural materials in the industrial revolution and electronic materials in the information revolution can be formulated as a guide to long-term expectations in ULSI.

The dominant chemical element of the industrial revolution has been iron (Fe). Silicon occupies a similar position in the information revo-

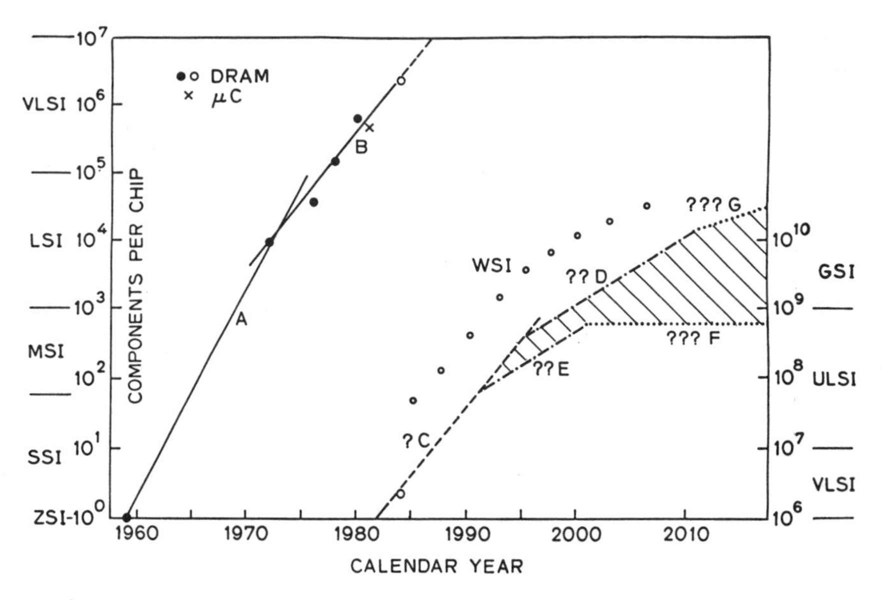

FIGURE 6 Components per chip versus calendar year. Segments A and B are history. Segments C, D, E, F, and G are projections. E F assumes 0.75-micron limit on optical lithography and 0.5-micron limit on IGFET minimum feature size. D G assumes 0.50-micron limit on optical lithography and 0.25-micron limit on IGFET minimum feature size.

lution.[27] An Fe/Si analogy can be extended hierarchically to compounds, components, and systems. By alloying carbon (C) and other elements such as chromium (Cr) and nickel (Ni) with Fe, special-purpose steel ingots can be cast. In analogous fashion, boron (B), phosphorus (P), or antimony (Sb) dopant is added to Si to prepare single-crystal ingots for various discrete devices or integrated circuits. Engine blocks and pistons are examples of specialized steel components corresponding to IGFET and bipolar integrated circuits intended for different purposes. Finally, automobiles, skyscrapers, and railroads represent the steel systems of the industrial revolution as portable computers, mainframes, and communications networks mark the silicon systems of the information revolution.

The principal technical figure of merit of structural materials is their strength-to-weight ratio. For electronic materials and devices, the reciprocal power-delay product is the most useful figure of merit. In applications such as high-speed aircraft where the strength-to-weight ratio of steel has been found wanting, aluminum (Al) has been a valuable replacement. For high-speed electronic circuits where Si devices are too slow, GaAs technology has prevailed, thus suggesting a second structural/electronic materials analogue. In the most demanding applications, such as jet engines, where the largest possible strength-to-weight ratios are required at very high temperatures, aircraft engine designers have selected titanium (Ti) as the appropriate structural material. A corresponding position is occupied by superconductive devices that offer the largest reciprocal power-delay product of all currently known electronic technologies. Moreover, the impact of superconductive devices on the information revolution is yet to come.

For several decades various types of plastics have served remarkably well as low-cost, lightweight, and in many instances superior-performing replacements for structural metal alloys. Current research to develop semiconductor-on-insulator materials (particularly Si) for integrated electronics is aimed at an analogous set of targets that again represent potential future advances. Finally, the most exotic or largest strength-to-weight ratio structural materials now in use are composites in which crystalline fibers of C, B, or glass, for example, are embedded in a host or binding material. Future multimaterial wafers such as Si-Ge-GaAs-GaAlAs structures, perhaps fabricated by molecular beam epitaxy, can be conceived as the counterparts of structural composites. Extension of this set of analogies will be left to the imagination of the reader (e.g., consider germanium [Ge]).

Annual production in tons of steel, aluminum, titanium, and plastics in the United States since 1860 is summarized in Figure 7. The figure also illustrates estimated annual world and U.S. production of single-crystal silicon and world production of all other single-crystal electronic materials in equivalent square inches of wafers. It is interesting to observe that the rate of steel production in the United States doubled every two years from 1860 through about 1910. For more than the past decade Si has followed the same trend. Another remarkable feature of the data is that after more than a century of high-volume production, steel remains the dominant structural material. Moreover, the history of Al, Ti, and plastic suggests a myriad of exciting future material, device, and circuit advances to propel the information revolution. The possibility that the future course of electronic materials may follow the historic patterns of structural materials implies an intriguing set of long-term analogical limits on integrated electronics.

CONCLUSION

The rate of progress toward the hierarchy of limits governing integrated electronics was extremely rapid during the 1960s and early 1970s.

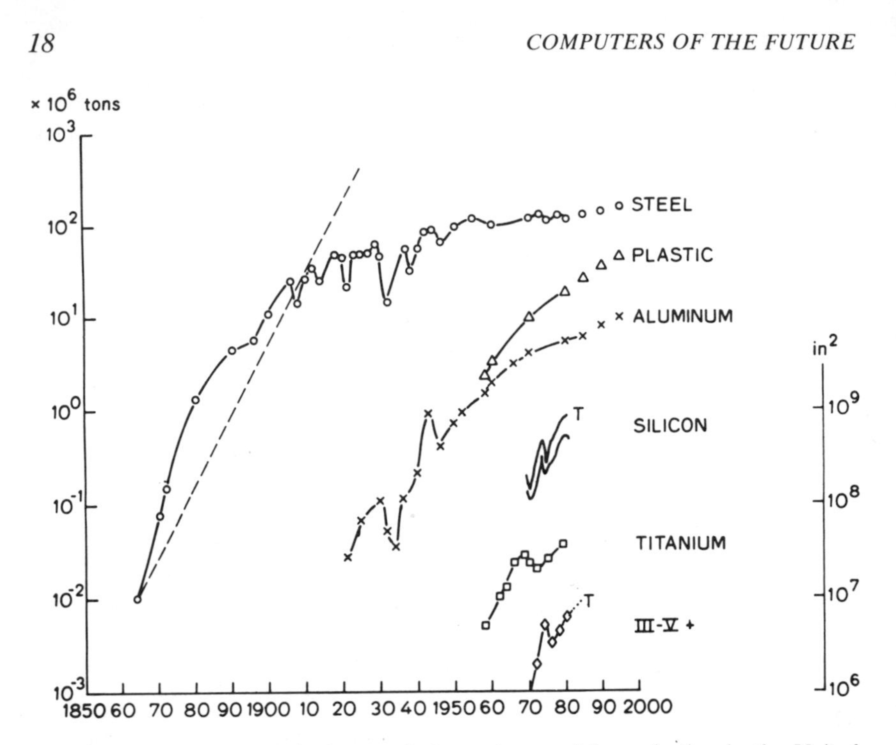

FIGURE 7 Structural and single-crystal electronic materials production in the United States and the world (T) versus calendar year.

A perceptible decrease in this rate has been observed since the mid-1970s and is projected to continue until the early 1990s. Theoretical limits on minimum feature size and practical limits on lithography and reliability predict a further rate reduction at that time. Nevertheless, as illustrated in Figure 6, chips incorporating several hundred million to a billion (i.e., gigascale integration, or GSI) components are anticipated by the year 2000.[28,29] Following this, analogical limits suggest a continued high level of utilization of silicon integrated electronics as well as rapid advances and volume applications of complementary technologies. Although these projections are quite encouraging, they are limited to extrapolation of already identified trends. The possibility of fundamentally new discoveries and inventions can only add to future prospects.

NOTES

1. G.E. Moore, "Progress in Digital Integrated Electronics," *Tech. Dig., 1975 IEEE IEDM*, pp. 11-13.
2. R.W. Keyes, "Physical Limits in Digital Electronics," *Proc. IEEE*, *63*:740-767, May 1975.
3. R.W. Dennard et al., "Design of Ion Implanted MOSFET's With Very Small Physical Dimensions," *IEEE JSSC*, *SC-9*:256-268, Oct. 1974.
4. VLSI Lab, Texas Instruments Inc., "Technology and Design Challenges of MOS VLSI," *IEEE JSSC*, *SC-17*:442-448, June 1982.
5. W. Fichtner et al., "0.15 μm Channel-Length MOSFETs Fabricated Using E-Beam Lithography," *IEEE ED Lett.*, *EDL-3*:412-414, Dec. 1982.
6. G.E. Smith, "Sub-micron NMOS Technology for High Speed VLSI," *1983 Symp. on VLSI Tech., Digest*, pp. 102-103, Sept. 1983.
7. H. Ihantola and J. Moll, "Design Theory of a Surface Field Effect Transistor," *Solid-State Electron.*, *7*:423-430, 1964.
8. R.M. Swanson et al., "Ion-Implanted Complementary MOS Transistors in Low-voltage Circuits," *IEEE JSSC*, *SC-7*:146-153, Apr. 1972.
9. L.D. Yau, "A Simple Theory to Predict the Threshold Voltage of Short-Channel IGFET's," *Solid-State Electron.*, *17*:1059-1063, 1974.
10. K.N. Ratnakumar et al., "Short-Channel MOST Threshold Voltage Model," *IEEE JSSC*, *SC-17*:937-947, Oct. 1982.
11. J.R. Pfiester et al., "Performance Limits of NMOS and CMOS," *IEEE ISSCC Tech. Dig.*, pp. 158-159, Feb. 1984.
12. H.I. Hanafi et al., "An Accurate and Simple MOSFET Model for Computer-Aided Design," *IEEE JSSC*, *SC-17*:882-891, Oct. 1982.
13. J.Y.W. Seto, "The Electrical Properties of Polycrystalline Silicon Films," *J. Appl. Phys.*, *46*:5245-5254, 1975.
14. N.C.C. Lu et al., "Scaling Limitations of Monolithic Polycrystalline-Silicon Resistors in VLSI Static RAM's and Logic," *IEEE JSSC*, *SC-17*:312-320, Apr. 1982.
15. N.C.C. Lu et al., "High Field Conduction Mechanisms in Polycrystalline Si Resistors," *Tech. Dig., 1982 IEEE IEDM*, pp. 781-785.
16. T. Kamins, "MOS Transistors in Beam Recrystallized Polysilicon," *Tech. Dig., 1982 IEEE IEDM*, pp. 420-424.
17. J.F. Gibbons et al., "A Folding Principle for Generating 3-D MOSFET's in Beam-Recrystallized Polysilicon Films," *Tech. Dig., 1982 IEEE IEDM*, pp. 111-114.
18. K.C. Saraswat et al., "Effect of Scaling of Interconnections on the Time Delay of VLSI Circuits," *IEEE JSSC*, *SC-17*:275-280, Apr. 1982.
19. J.D. Meindl et al., "Circuit Scaling Limits for Ultra-Large Scale Integration," *Digest, 1981, IEEE ISSCC*, pp. 72-73.
20. P.E. Cottrell et al., "Multidimensional Simulation of VLSI Wiring Capacitance," *Tech. Dig., 1982 IEEE IEDM*, pp. 548-551.
21. H.B. Bakoglu et al., "Optimal Interconnect Circuits for VLSI," *IEEE ISSCC Tech. Dig.*, pp. 164-165, Feb. 1984.
22. A.N. Broers, "Resolution, Overlay and Field Size for Lithography Systems," *IEEE ED*, *ED-28*:1268-1281, Nov. 1981.
23. J.H. Bruning, "Optical Imaging for Microfabrication," *Semicond. Int.*, pp. 137-156, Apr. 1981.
24. D.B. Tuckerman et al., "High Performance Heat Sinking for VLSI," *IEEE ED Lett.*, *EDL-2*:126-129, May 1981.
25. J.P. Martino, ed., *An Introduction to Technological Forecasting*, Gordon & Breach, New York, pp. 13-25, 1972.

26. O.G. Folberth et al., "The Fundamental Limitations of Digital Semiconductor Technology," *Microelectron. J.*, *9*:33-41, 1979.
27. J.D. Meindl, "VLSI and Beyond," Keynote address at Microelectronica 1982, Eindhoven, The Netherlands, Jan. 1982.
28. R.W. Keyes, "The Evolution of Digital Electronics Towards VLSI," *IEEE ED*, *ED-26*:271-279, Apr. 1979.
29. J.D. Meindl, "Theoretical, Practical and Analogical Limits in ULSI," *Tech. Dig.*, *1983 IEEE IEDM*, pp. 8-13.

The Evolution of Computer Systems

HERBERT SCHORR

For large computers since 1955, the cost, in current dollars, has dropped by a factor of more than 200, while processing speed has increased by a factor of nearly 400. It is expected that the evolution will continue for systems and that large-machine performance will increase 8 times in the next decade. This growth will come both from the continued improvements in technology that will reduce cycle times and from improvements in the design of processors that will reduce the number of cycles required per instruction.

The amount of information stored in electronic form is increasing at some 40 percent a year. The management of massive data bases to which many people need quick access requires a lot of computational power, and once these data bases are on-line, business growth typically drives the demand for workstation access to them even higher.

Today's state-of-the-art IBM 3380 stores 6,000 times more information per area of disk surface—at 150 times less cost per character—than did the first disk storage unit announced by IBM in 1956. Thin-film read/write head technology pioneered by IBM makes it possible to transfer data at up to 3 million characters per second. Even so, today's faster computers are restrained by the time it takes to locate information, so in this area new inventions and memory hierarchies will be required.

The need of IBM's large commercial customers for computational power is growing at about 40 percent a year, and this need will continue to be responded to by multiple-machine environments combining uniprocessors of ever-increasing power. Machines will be combined in a variety of ways: closely coupled, loosely coupled, through shared resources, and over high-speed buses. However, the degrees of parallelism are likely to be moderate in that only 16 or 32 main processors will be combined in single complexes.

Large-scale integration has made possible processors of moderate capability on a single chip. The performance of these microprocessors is growing steadily. Data-flow widths have increased from 4 to 32 bits. Some new microprocessors have on-chip caches, and cycle times can be expected to decrease significantly from the current level of 200 to 400 nanoseconds. Increases in performance of 25 to 50 times can be expected in the coming 10 years.

Microprocessors offer the possibility of alternative architectures to today's configurations where most processing is done by the central host. These alternatives involve moving function from the host out to the attached devices in the form of intelligent workstations, printer servers, and file servers. Custom and semicustom very large scale integration (VLSI) is also being used to migrate function from software to hardware, to implement specialized functions (such as data compression), and to reduce the amount of random logic in small systems.

The availability of inexpensive, high-performance microprocessors has been responsible for the explosive growth in personal systems today, and this trend is expected to continue. The performance of these microprocessors will enable workstations to handle multimedia data types, such as image and digitized voice.

It should be possible in the future to build computers that accept and "understand" natural human speech and respond in a way that is appropriate, useful, and intelligible. Machine recognition of discrete-word speech by a speaker known to the machine is practical today. Continuous speech of high complexity has been understood by a computer at IBM with greater than 90 percent accuracy, but the required computational power is still uneconomically large. Increased performance, specialized hardware, and improved techniques will enable speech-recognition functions to be commonly integrated into workstations in the coming decade.

The growing trend toward personal computing parallels a phenomenon that took place with general-purpose minicomputers in the mid-1970s. The so-called minirevolution helped provide much-needed productivity in a hurry, but it eventually resulted in information bottlenecks throughout many enterprises. Because there was no commonality or central planning, information collected or produced on local processors was not compatible with information collected or produced in the rest of the organization. To avoid a repetition of that difficulty, serious thought must be given to interconnecting an organization's information-handling systems—whether word processors, host-attached terminals, or personal computers—with each other and with current host systems.

The term *systems interconnection* denotes the distribution of function and data among different systems within a network. Systems interconnections comprise several key components, including the network transport facility; a set of higher-level protocols for session and application; and methods for locating, requesting, and managing distributed resources. The goals of systems interconnection include resource and data sharing, modular systems growth, high availability, and growth that reflects the distributed nature of organizational structure. An important model is local clusters of intelligent workstations and servers on local-area networks (LANs) for the purpose of resource sharing. These systems consist of intelligent workstations accessing common storage and printing devices called file servers and printer servers. A file server provides disk space for stations that are not equipped with hard disks and allows storage, retrieval, and sharing of files. One main objective is to allow users at intelligent workstations to access large, shared-host data bases as well as expensive, host-based resources such as high-quality printers.

As computers become a daily part of more people's lives, attention is shifting from the central electronics complex to systems issues: programming, human factors, communications, and applications. The programming and the end-user interface are in many ways the limiting factors that determine how broadly the productivity benefits of information technology can be realized.

Over the years computer users have found that an increasing fraction of their total cost lies in programming. This includes both the programming supplied by the computer manufacturer or other vendors and the programming done by users themselves as they evolve from batch processing to telecommunications-oriented systems, and finally to distributed complex systems. More circuits and hardware will pay for themselves if they make this programming job more productive.

During the 1980s, support of the world's 30 million or so office principals will become a major focus of effort. Eventually office systems will have to be able to merge today's separately handled data, text, voice, and image information into integrated electronic documents that can be communicated, retrieved, and otherwise dealt with, without the user's being conscious of the form of the data when stored. Artificial distinctions between data, word, and image processing will gradually disappear as document distribution, filing, and retrieval systems for all kinds of documents evolve.

What about the university of the future? Universities across the land are entering an era of information systems experimentation. One prototype is beginning to emerge at Carnegie-Mellon University (CMU) in Pittsburgh where IBM and the university have just begun the joint development of a unique, personal-computing network.

One of CMU's primary goals is the full integration of computing into undergraduate and graduate education. Eventually, it is expected that all CMU students, faculty, researchers, and professional staff will have access to personal-computer workstations that are effectively 20 to 100 times more powerful than current home computers. Each will also have access to shared, central data bases through a high-speed, local-area network. By 1986 several thousand of these personal workstations should be in place.

The planned configuration will consist of four system elements: the workstations, local-area networks or clusters, a backbone communications network with gateways and bridges to the other elements, and a central computing and data base facility. In addition to providing traditional computing capabilities, the personal workstations will enable users to work on several projects simultaneously, to create drawings and diagrams, and to see these exactly as they will be printed; the workstations will have provision for audio input and output.

During the 1980s work on communications system architecture will have to focus on the three fronts discussed below: (1) the geographically distributed communications network, (2) communications within the establishment, and (3) the gateways between them. Modern data networks will adapt themselves to the needs of the people instead of making the organization conform to the system's structure.

The U.S. Department of Defense made a greater contribution than it knew in establishing ARPANET, a network linking research groups in U.S. universities, in the early 1970s. The original idea was to permit sharing of computers that were not equally loaded. It turned out that this network was really used to discover the power of electronic mail and to document creation and distribution. This led to a new capability for collaboration by scientists thousands of miles apart. IBM has also had this experience on a much larger scale with an open-ended, peer-connected network called VNET, which contains more than 1,000 host processor nodes in hundreds of cities in 18 countries. VNET grew "bottom-up," so to speak, when two laboratories working on a joint project needed to exchange data. Soon other related sites were added, and the network grew until virtually all of IBM's scientific engineering locations worldwide are part of it. It is still growing. About 50 university computers are also hooked up, using this capability under the name BITNET (the acronym means "Because It's There"). Such networks have an almost unique ability not only to improve the effectiveness of working groups, but to foster subcommunities of interest that the host organization is not even aware of.

For communications among establishments, computers and their users are already starting to benefit from the many public digital communications services and data networks now offered or planned and from communications satellites through which data can flow at the same speed at which the computers themselves operate. This external telecommunications environment will, no doubt, continue to evolve, and agreement on and development of new standard interfaces for voice and data by common carriers around the world is another virtual certainty.

But the main technical thrust of the 1980s lies elsewhere—within the establishment, where local-area networks (LANs), the private branch exchange (PBX), and host attachments are the key elements in this rapidly changing environment.

An LAN can be defined as any information transportation system that provides for the high-speed connection between users within a single building or a campus complex, through a common wiring system, a common communications adapter, a common access protocol to allow connection between users, and common shared resources, such as the largest files and the most powerful printers. Several LAN approaches are in contention, including IBM's token ring architecture.

Sorting out the technical merits and application domains of the token ring, contention bus, and competing LAN approaches will be a major activity of the 1980s. However that debate turns out, clearly, local-area networks must be integrated into network architectures. And that is driving the further evolution of IBM's System Networking Architecture (SNA) from its original hierarchical, large-host orientation toward one that increasingly fosters peer-to-peer communications.

The typical 1980s PBX will have a significant and growing role. It can be expected to provide not only traditional PBX functions—such as attachment and control of conventional telephones, and data devices using modems—but advanced voice capabilities such as speech messaging. In addition to attachment of analog telephones and data devices with modems, there will likely be digital telephone attachments and direct digital attachment of terminals to PBXs.

A facility that will continue to have a significant role in establishment and office systems of the future is the host-attached controller. It will evolve to include local-area network functions for workstations and terminals while offering the facilities to a central host.

All three of these intraestablishment media will be required to support the full range of applications that will evolve in future establishment systems. Many customers will demand capabilities that can be provided only by having more than one. This will increase the need for gateways that can bridge between both these different local forms and between local and wide-area communications systems.

As the success of data base/data communications applications spurs the rapid growth of intelligent, programmable workstations, the personal computer and the data terminal are beginning to merge. The wave of the future may well be something that looks very much like an easy-to-program, more powerful version of a personal computer and that is networked both to similar workstations and to shared input/output devices as well as to shared data on larger host systems.

Software and Unorthodox Architectures: Where Are We Headed?

MICHAEL L. DERTOUZOS

SOFTWARE

Can you construct a 10-mile-long wooden bridge? The answer is not very relevant. What is relevant is that if this question is posed to 100 civil engineers, they will answer either yes or no fairly consistently.

Now for a question at the cutting edge of computer science research: Can you mimic an airline office by a computer program so that when someone types in a query that would normally be placed by telephone—"Do they serve meals on the flight from here to Atlanta?" or "What time does the flight leave?"—the automated system responds on that person's terminal with the kind of answer that would be given over the telephone? Can you construct a system that will answer 80 percent of these questions correctly? I have asked this question of many people working on the leading edge of computer science research, and their typical answer is "Give us $2 million and about two years." While I have not carried the experiment further, I know that if I were to pursue it, in two years the answer would be either "Here it is, and it's answering correctly 75 percent of the queries" or "Give us another $2 million and another two years." The point here is that with the exception of computer science theory and compiler design, there is very little science in this field, which in my opinion is still largely experimental in nature. The cutting edge of computer science is very much like alchemy in the old days—mix two liquids and if there is an explosion take notes. Leaving some room for poetic license, that is the approximate situation with most of our advanced programs and with many of our avant-garde hardware architectures, as will be seen.

As a result of this absence of "laws" in the world of software, the programmer is the lawmaker. That, in a nutshell, is the power and attraction of programming: to set good laws and to build complex structures based on these laws. For this reason programmers are extremely reluctant to "buy" each other's laws; each one wants to be the lawmaker. That is also why young people become millionaires by starting programming companies—their fresh minds are not burdened by the excess baggage of experience, and they can accordingly thrive in creating and adhering to new laws.

What can educators really teach to a young aspiring programmer? Unfortunately, very little. Educators can issue generalities and platitudes, such as, "Design your programs in nice modules, isolate them, make each module neat and tidy and describe its inputs and outputs." This is an exaggeration, of course, but there is very little about programming that can be taught explicitly. Educators do ask young programmers to read other people's programs and to apprentice by watching experienced programmers at work. The effectiveness of these approaches again suggests that the similarity to alchemy or even to sculpture may not be too farfetched.

Traditional software generation cannot be substantially improved with the techniques that are in hand today or that have been developed over the last 20 years. While the hardware cost/performance improves at the rate of roughly one decade per decade, the corresponding software improvement is by comparison insignificant: During the last 10 years software productivity measured as lines of program produced per programmer per year has increased by about 50 percent.

Are there any prospects for improving software productivity? I will discuss a few. *Structured programming* is a technique that is most effective if one deals with very small programs, often three or four lines long. I am reminded of the pioneer of structured programming, the famous Professor Dijkstra, who spent an hour explaining to a large audience at MIT how a three-line program could be verified. One of our students asked: "But how would you apply this to a chess-playing program?" And Dijkstra, in his inimitable style, said: "One should never write such complicated programs." More seriously, I believe that structured programming has been helpful, but we seem to have exhausted its benefits.

Automatic programming was a technique attempted, without much success, during the last 15 years. The intent was to describe something like an inventory control process at a very high level and then to use a gigantic compiler to generate the high-level programs (e.g., in COBOL) that one would normally write for inventory control. This approach did not work well for reasons of complexity and variance among different users' needs.

Functional programming is a recent approach that tries to build programs in the form of mathematical functions that have inputs and outputs and, significantly, no side effects. Different functional programs can be combined in the same way that complex mathematical functions can be composed out of simpler ones. It is too early to predict how well this approach will work, and it is difficult at this time to bend into the functional mold traditional programs used, for example, in data base management, where side effects are the governing mechanisms.

There is currently great hope that *expert systems* will be useful in programming; this would mean that an "expert assistant"—a program—would help us program more effectively. Unfortunately it is too early for a meaningful assessment of this prospect.

My own overall assessment about the future success of programming productivity improvements is rather bleak. But let me ask a provocative question: Why do we need programming?

To go back 20 to 25 years—in the days of batch computing people programmed because there was a very expensive resource called the computer that cost several million dollars to acquire. They waited in line for hours if not days to get their cards processed into results, only to modify their programs against that "last bug" so that they could stand again in the waiting line.

Later, in the late 1960s, time-sharing came along. At the beginning at least, the reasons for developing and using time-shared systems were the same. The same very expensive resource could now be shared by many people in round-robin fashion so that each user did not, it was hoped, notice the delay. Soon thereafter, time-sharing pioneers changed their position and said that cost-sharing was no longer a valid reason for these systems. Instead, attention was focused on the utility aspects and the services offered by those systems. And that, indeed, turned out to be one of the great truths about the ultimate usefulness of time-shared systems.

Now, with computer costs dropping dramatically, the personal machine that I have next to my desk costs $5,000 and does more than the $2.5-million time-shared machine that we had in our lab 20 years ago. To the extent that this is now the case, why is programming needed anymore? Certainly not for cost-sharing. It is needed for *application flexibility*, i.e., to carry out different applications on the same piece of equipment.

Let me now ask a second provocative question. We have said that while the hardware is improving at the rate of 30 percent per year, software productivity is essentially standing still—hence we are confronted with a "great software problem." The question is simply this: Is there really a software problem? In my opinion the answer is no for this reason: Because software involves paper and pencil and not the traditionally difficult tasks of putting hardware together into complex physical systems, there is a tendency to assume that software ought to be easier. This is even reflected in the legal differentiation made between software and hardware: one is patentable while the other is "copyrightable."

Compare, however, the design of the airline office program mentioned earlier with the design of a complex physical system like a jumbo jet—and not the routine twentieth design of a jumbo jet, but the original design of such an airplane. This comparison deals with two objects of roughly the same complexity in terms of the elements that they contain. And while people might laugh about the airline office program taking several years to design, they would readily agree that such a long design time would be reasonable and proper in the case of the jet plane.

In my experience there is neither precedent nor prospect for achieving *dramatic* economies in the design of complex systems, especially in the case of first designs, which are typical of novel software systems. In these cases creativity plays an important role since one is trying to break new ground. There is no reason to believe that in the field of programming this endeavor ought to be easier than in any other discipline that depends heavily on creative design.

Let me close this section on software prospects with some good news. There is ample precedent for economizing through the mass production process, e.g., by spreading the very expensive cost of designing a complex product to the people who buy it.

Accordingly, I would suggest that one of the biggest forces before future software developments is the transition of the software development process from the artisan to the mass production era. In particular, two significant categories of products can be expected to emerge in the next decade.

The first category can be called *hidden computers*, because it deals with "appliances" performing a useful function. These products will not be thought of as computers any more than cars are thought of as collections of carburetors and other components, or refrigerators as electric motors. Each hidden computer is especially programmed for one application, not changing its program, and providing immediate utility to its user. Examples include the drugstore machine, the personal memo pad, the automobile safety package, the automobile convenience package, the fuel control package, the automobile maintenance package, and so on.

The second emerging category of software products is *mass-manufactured applications* in the form of diskettes, which in time will become more powerful and more useful. In both categories the large costs incurred in the design and production of these items become very small when spread to hundreds of thousands or even millions of users.

Finally, I do expect that more intelligent programs will make life easier for all of us. When I am told that a certain program or machine has a friendly user interface I get a bit worried, because I find that I need 40 or 50 hours to become truly acquainted with a new machine, and I am a computer professional. Such a relationship may be friendly compared with a life-threatening situation, but it is not a relationship we should have with something next to our desk. I believe that the only way to achieve true friendliness with computers is by increasing the intelligence of the programs that they contain. Only then will machines begin to understand what we are trying to do in the same sense that friends understand what we mean when we communicate with them.

FORTHCOMING SYSTEM ARCHITECTURES

The dominant theme in forthcoming machine architectures will be the *myria-processor* system. In Greek, *myria* means 10,000. That number may be close to what will be used in practice, at least in the next 15 to 20 years. Look, then, for future computer systems that use hundreds or thousands of processors.

Myria-processors will be used in two areas. The first involves the so-called *geographically distributed systems*. This architecture is evolving because people who generate and collect data are geographically distributed. Moreover, organizations have geographically distributed offices and personnel who need to communicate with one another. And, of course, there is a great techno-economic opportunity along with these needs in that people can now afford to buy computers and the communications products and services that interconnect them.

The second area of myria-processor revolution involves *tightly coupled multiprocessor* systems. Here I envision again hundreds or thousands of processors in one very big box, devoted to one task, for example, recognizing the face of one human being among a few thousand or trying to comprehend human speech. As in the case of distributed systems, a sizable techno-economic opportunity is driving us in this direction, with the very low cost of individual processors making such large systems affordable.

Geographically Distributed Systems

Consider 1,000 autonomous machines that are geographically distributed—autonomous because each machine can support the desires of its owner in editing text, handling messages, and running application programs, all independently of what other people are doing on their computers. In addition to this autonomy, however, each computer must cooperate with the other computers in the system, whether they are 1,000 miles away or upstairs, in order to make possible common applications, e.g., planning, putting together a manual, or trying to figure out why there was a power failure in New York City.

The electronic interconnections among computers, which are a combination of local-area networks and long-distance terrestrial or satellite networks, are fairly straightforward at the hardware level and present no substantive long-range problems. Large companies that have begun to establish networks of distributed systems are approaching these systems from an era in which dumb terminals became progressively more intelligent in the presence of a central powerful program that controls everything in the system. It is now becoming clear that these interconnected machines need to be autonomous and that there can be no central "executive." The reason for this decentralization lies in the desire to establish arbitrarily large systems: A centrally directed system has an upper size limit of perhaps 50 to 100 nodes for the same reason that an individual can only converse simultaneously with a small, limited number of people. Achieving truly decentralized systems will require the development of some form of common currency—or, if you wish, simple English—for computers so that these machines can "understand" each other for the purpose of common applications. This requirement is at least philosophically analogous to the communication means in humans tribes: Each individual is autonomous, yet the aggregate can embark on common activities because they can understand one another. Ultimately these computer tribes will form an "information marketplace," i.e., a free market dedicated to the purchase and sale of information and informational labor.

Now let me pose a challenge to computer professionals: Can you construct a 1,000-computer distributed system that at minimum has the behavior and performance of a traditional, centralized, time-shared system that services perhaps 50 users in round-robin fashion? The choice of the number 1,000 is of course arbitrary. If this goal can be achieved for 1,000 machines, I would immediately ask if it can be done for 10,000 or even a million computers. In short, is there a *scalable* system architecture that permits the kinds of services and the sharing of data that transpire in time-shared communities? I have posed this challenge in terms of time-shared systems because there is experience in that area, and it is known that if this behavior and performance can be extended to arbitrarily many users the result will have proven utility.

This question has not yet been answered. It seems, however, that it must be answered one way or another before distributed computer systems become truly useful.

Let us look, for comparison purposes, at an existing network called the ARPANET, which spans several hundred computers from Hawaii to West Germany. In principle and in practice it is possible to log into any computer on this network and to use that computer from a remote location. Yet, hardly anyone does this. Instead, people use this network as a very sophisticated message-exchange medium, which has proven its utility for that purpose. The reason that remote use of computers does not take place is that dialing into another computer generally means being confronted with an entirely different system that speaks its own language and has its own conventions, which are either unknown to the person dialing in or are costly to learn. Hence, except for the great young hackers who enjoy exploring remote computers, most people have very little use for such an unknown and remote resource.

Regardless of the success of devising effective decentralized and distributed computer systems, individual computer "nodes" are likely to continue their rapid evolution. The dominant trend that I see is toward more powerful personal computation. In this regard people often ask if they really need in their personal computers more memory than 128,000 characters, half a million characters, or at most 1 million characters (1 megabyte). The answer is an easy yes. If we focus on greater program intelligence, which is the key to making these machines truly useful to us, then we shall discover that a forefront-research intelligent program occupies today 3 or 4 megabytes and, incidentally, takes about 30 to 70 man-years to develop. So having a personal computer that is truly friendly and that tries to comprehend some free-format English as well as to provide certain intelligent services involves the equivalent of 4 or 5 of today's forefront intelligent programs, i.e., some 15 to 20 megabytes of primary memory.

To conclude, I believe that success in distributed systems will depend on our ability to address effectively the following three challenges. First, such systems must achieve an acceptable level of overall reliability in spite of local failures. Currently computer systems are like Swiss watches.

Reach into them with a spoon and the entire edifice fails, catastrophically. It is essential to make distributed systems work more like human tribes—if one or more nodes collapse, the tribe goes on functioning properly.

The second and perhaps biggest challenge before us is to discover a minimal semantic base, a minimal form of "computer English" that is understood by all machines and that simultaneously maximizes local autonomy and application cohesiveness.

The third challenge involves the assurance that the information carried on distributed systems is protected and that the signatories of messages are indeed the purported agents and not impostors.

Multiprocessor Systems

The second area of myria-processor systems might be called *computation by the yard*, i.e., if you like 1 yard of computing you buy a 100-processor system, and if you like 2 yards you buy a 200-processor machine. An example of a multiprocessor system is a 1,000-processor machine that converts human speech to typed text. Other applications include the solution of large partial differential equations, signal processing, and weather forecasting. But the most interesting applications are in *sensory computing*, meaning automated eyes and ears, and in more *intelligent computing*.

The force driving us toward multiprocessor systems is their cost. If 1,000 small processors can be harnessed into one computing machine, then that machine will have 10 times the processing power of today's fastest computer, at 1 percent of the latter's cost.

This harnessing of numerous processors in one machine is being attempted in several ways. One way is the *data-flow* approach, which consists of (1) a communication network that has perhaps 1,000 inputs and 1,000 outputs, and (2) 1,000 processing elements. In this architecture the results issued by each computer enter this network carrying a little flag that establishes their destination. Thus, if the flag says "39," these results will exit the network into the thirty-ninth computer where they will be processed into new results with new flags.

The programming of such a machine is entirely different from the programming of today's Von Neumann-type machines, because many operations are carried out simultaneously rather than sequentially.

The challenges in trying to realize these multiprocessor systems are as follows. First, such systems must be linearly or at least quasi-linearly scalable—this is the computation-by-the-yard notion. Second, as in the case of distributive systems, these multiprocessor machines must work correctly in spite of local failures, because in a system with 1,000 processors there will always be several processors that do not work correctly. Finally, there must be an associated programming environment so that these machines can be effectively programmed. Moreover, such a programming environment should be "tuned" to the hardware architecture to ensure good performance under the target applications.

The following question is often raised in discussions of multiprocessor architectures: It is conceivable that we can put together 1,000 or 10,000 processors, but how are such systems going to extract the parallelism inherent in today's large FORTRAN programs? My answer is that this is not possible, any more than it is possible to automatically extend procedures for managing one individual to procedures for managing a factory or a company of 1,000 people. In both cases the techniques are entirely different. Hence, we must forget about automatically unfolding the parallelism in today's programs. Instead, it is necessary to learn how to harness these "warehouses" full of computers by explicitly programming and organizing them in the same sense that a complex shoe-manufacturing company is planned and tuned today.

Finally, I think that following our experience with single-processor machines there will be *special-purpose* multiprocessor systems dedicated to specific applications that will always carry out these applications without the need or ability to be reprogrammed for other applications.

SUMMARY AND CONCLUSION

In the software domain, programs will move from the artisan to the mass-manufacturing era, and they will manifest themselves progressively more as hidden computers or mass-manufactured applications on diskettes. Users will adapt to these programs rather than the other way around as has been the case in the past. Programs will become more intelligent and will move into personal computers, distributed systems, and multiprocessor systems where they will perform a variety of service-oriented functions, including the sensory area. Programming productivity will not increase dramatically. Instead, the continuing large program design costs will be absorbed in the individual cost of mass-produced programs.

Novel system architectures will abound under the myria-processor theme in two principal categories: (1) geographically distributed, loosely coupled systems that behave, like tribes, with autonomy and application cohesiveness and that will form an information marketplace; and (2) multiprocessor tightly coupled systems, each dedicated to one application, which are expected to open new doors in sensory and intelligent computing.

Part II

New Frontiers in Biotechnology

Introduction

RALPH HARDY

To provide an overview of what is called the new biotechnology, we have assembled three of the leaders in this field from bioengineering, health care, and agriculture. But first a brief introduction.

The reality of biotechnology is strikingly shown in Figure 1. Genes that direct the synthesis of the animal growth hormone were injected into the embryo of the rodent pictured on the left-hand side of the figure. The addition of genetic material by the new biotechnology led to a very substantial change—up to a doubled size in this organism. Clearly, then, there is a reality to biotechnology.

As discussed by Charles L. Cooney in the first of the three papers that follow, processes for many of the biotechnology applications require engineering input. Some believe, in fact, that the process aspect of this research may be more demanding than is the genetic manipulation aspect.

The reality of biotechnology in agriculture is illustrated by the knowledge that the change of a single amino acid in a protein that occurs in the membrane of the chloroplast, the light-gathering area of a plant, led to resistance to a major herbicide called atrazine. In fact, rapeseed with this resistant gene is now being aggressively used by farmers on the North American continent. Charles J. Arntzen provides an overview of this and other aspects of biotechnology in agriculture in the second paper.

The reality of biotechnology in health care is demonstrated by the clinical use begun in 1982 of human insulin produced in a fermentation

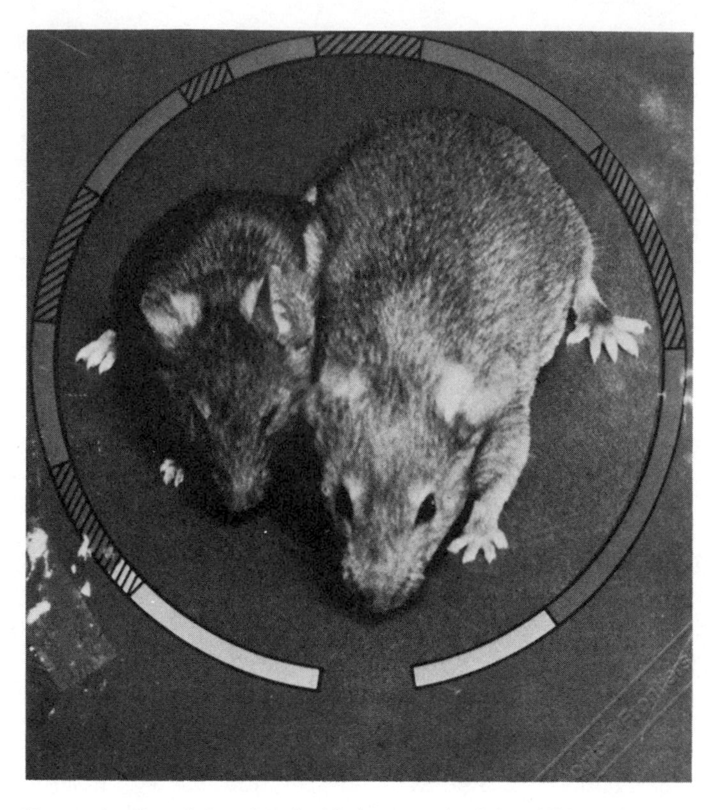

FIGURE 1 Size of organism doubled as a result of the addition of genetic material by the new biotechnology. Reprinted, with permission, from *Science*, Vol. 222, No. 4625, Nov. 18, 1983. © 1983 by American Association for the Advancement of Science.

vat in a microorganism into which had been inserted the genes for human insulin. J. Paul Burnett discusses this and other aspects of biotechnology in the health care area in the third paper.

The examples mentioned thus far clearly show that there is a reality to biotechnology. And clearly the fundamental, pervasive ability of this technology—the ability to manipulate in a directed sense the informational system, or genes, of all living cells—suggests that the reality of biotechnology will have impacts in a number of industries. These industries range from the nearer-term health care and agricultural areas through food, industrial chemicals, energy, forestry, pollution control, and mining, and possibly in time to bioelectronics.

However, such a pervasive capability has unfortunately led to many unrealistic expectations. It seems that countries without substantial biotechnological capability feel that they are not first-class countries, and the same thing has happened in many industries, among the public investors, and so forth. In the first six months of 1983, for example, the public committed about $250 million to new stock offerings to small biotechnology companies, almost all of which have no product now and which probably will not have a product for some time. Limited R&D partnerships in this area are beginning to become very popular.

What we would like to do here, then, is to provide an analysis of what the reality is in this exciting field with the tremendous potential. Clearly the new biotechnology is in its very early stages.

Biochemical Engineering Solutions to Biotechnological Problems

CHARLES L. COONEY

Biotechnology is often defined as the integration of biochemistry, microbiology, and process technology for the purpose of manufacturing or environmental management. Thus, biotechnology is a broad area that encompasses much more than genetic engineering and hybridoma technology. Furthermore, the translation of scientific discovery to commercial reality requires tremendous skills in process development using both existing and new technology, and, most importantly, it requires the integration of basic biological sciences with chemical process as well as electronic and mechanical engineering.

Several seemingly unrelated developments that have occurred simultaneously over the past several years have stimulated activity in biotechnology. First, the discoveries in genetic engineering that permit one to move DNA between different organisms and to amplify its expression into proteins allow scientists to do things that could not be done before. Second, volatile pricing in traditional feedstocks used in the chemical process industry (CPI) has catalyzed interest in the use of biotechnology to access inexpensive alternative and renewable raw materials for use in chemicals manufacturing. Third, some of the key patents protecting some major products of the pharmaceutical industry either have expired or will expire soon. As a consequence there is need to develop new technology or to improve existing technology used for manufacturing these products. Fourth, changing consumer demands, with greater concern for safety, convenience, and environmental impact, require new or improved products. Insulin is one example—highly purified human insulin rather than traditional products is desired. Sweeteners are another example; for instance, aspartame is replacing some of the traditional and less desirable artificial sweeteners. Fifth, biotechnology provides opportunities for developing new products as well as for improving the manufacture of existing ones. New technology provides a vehicle for corporate growth by improving existing processes and developing new markets, as exemplified by companies in the business of building manufacturing plants or process equipment.

In the sections that follow, problems of the chemical process industry are discussed first. To help assess whether biotechnology can be used to address some of these problems, the second section examines the current biochemical process industry. The third section discusses the use of genetic engineering in addressing CPI problems. Biochemical process development is then discussed, and the concluding section enumerates problems in process development that need to be solved in order for the results of the molecular biologist to be translated to goods for the consumer.

CHEMICAL PROCESS INDUSTRY

Biotechnology can solve problems. That, after all, is what engineering is all about. In this regard it is interesting to examine the problems of the chemical process industry, which are summarized in the following list:

- High feedstock cost
- Volatile feedstock pricing
- Overcapacity
- Energy-intensive processes
- Environmental and safety concerns

There is a strong dependence on feedstock cost, which typically comprises 50 to 75 percent of the manufacturing cost for commodity products. The feedstock costs are quite volatile, and it is not clear where or how fast they are going to change, except that they will generally increase. Overcapacity in the industry, which now operates at about 70 percent of nameplate capacity, makes it difficult to implement new technology unless there are very large savings. A strong dependence on energy costs further increases the concern over petroleum feedstock pricing. Lastly, an increasing awareness of environmental and safety factors surrounding the operation of chemical manufacturing plants makes less hazardous processes attractive.

Can biotechnology solve some of these problems? Do these problems of the CPI represent opportunities for biotechnology? The answer is a definite yes. One important attribute of bioprocessing is the possibility of assessing renewable and possibly cheaper raw materials, which could be important to the CPI. Such feedstocks may be lower in cost and may be readily available in local situations. They may provide swing capacity in cases that involve not displacing an existing chemical industry but picking up growth or taking care of fluctuation in demand. In addition, there may be potential for less hazardous operation or less negative environmental impact.

BIOCHEMICAL PROCESS INDUSTRY

In order to assess whether or not one can use biotechnology to address some of these problems in the CPI, it is important to examine the current biochemical process industry (BPI), exemplified by processes for production of enzymes, antibiotics, amino acids, organic acid, and so forth, alongside of the CPI. How do the CPI and BPI compare and what are the differences?

Figure 1 plots selling price versus annual production volume for a wide variety of biochemical and chemical products (see Table 1). The correlation between the price in dollars per pound and production in tons per year is quite good over a wide dynamic range from 1 through

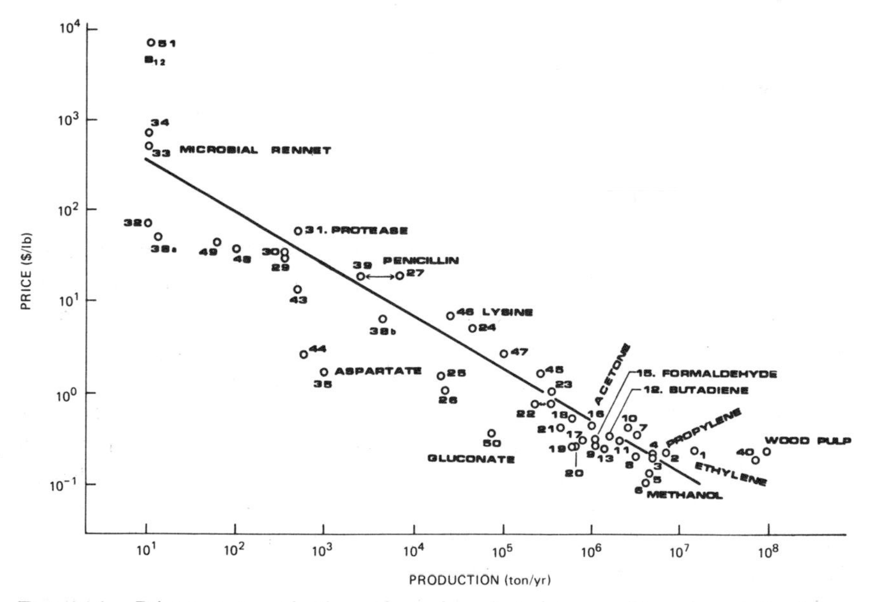

FIGURE 1 Price versus production volume for selected commodity and specialty chemicals and biochemicals (see accompanying Table 1 for listing of individual products).

TABLE 1 Price Versus Production Volume for Selected Commodity and Specialty Chemicals and Biochemicals (See accompanying Figure 1)

	Chemical or Biochemical	Production (1,000 tons/yr)	Price ($/lb)
1	Ethylene	15,000	0.25
2	Propylene	7,000	0.24
3	Toluene	5,000	0.21
4	Benzene	5,000	0.23
5	Ethylene dichloride	4,600	0.14
6	Methanol	4,200	0.11
7	Styrene	3,300	0.38
8	Xylene	3,200	0.21
9	Formaldehyde	1,100	0.29
10	Ethylene oxide	2,600	0.45
11	Ethylene glycol	2,100	0.33
12	Butadiene	1,600	0.38
13	Acetic acid	1,400	0.26
14	Phenol	1,300	0.36
15	Acetone	1,100	0.32
16	Pyopylene oxide	900	0.44
17	Isopropanol	800	0.32
18	Adipic acid	600	0.57
19	Ethanol (synthetic)	600	0.27
20	Ethanol (ferm)	700	0.27
21	Dextrose	430	0.45
22	Citric acid	300	0.80
23	Monosodium glutamate	350	1.10
24	Lysine	43	5.20
25	L-glutamate	20	1.80
26	Fructose	22	1.10
27	Penicillin	7	20.00
28	Glucose isomerase	1	25
29	Glucoamylase	0.35	30
30	Bacterial amylase	0.35	35
31	Bacterial protease	0.5	60
32	Fungal protease	0.0	70
33	Microbial rennet	0.0	500
34	Pectinase	0.1	700
35	Aspartic acid	1	1.75
36	D-HO-phenylglycine	—	13.6
37	D-phenylglycine	—	7.7
38a	D-calcium pantothenate	4.5	6.4
38b	D-calcium pantothenate	13	15
39	Penicillin	3.6	19
40	Palm oil	77,000	0.20
41	Cephalosporin	0.55	—
42	Tetracycline	3.4	—

(Continued)

TABLE 1
(Continued)

	Chemical or Biochemical	Production (1,000 tons/yr)	Price ($/lb)
43	L-arginine	0.5	13.6
44	L-aspartic acid	0.6	2.7
45	L-glutamine	270	1.8
46	L-lysine	25	7.3
47	D,L-methionine	100	2.9
48	L-phenylalanine	0.1	38
49	L-tryptophan	0.06	44
50	Gluconate	75	0.48
51	Vitamin B_{12}	0.0	7,000

NOTE: Numbers in left-hand column correspond to individual points on Figure 1.

100 million (10^8) tons per year, and from $0.10 to $10,000 per pound. It is interesting that the correlation holds for chemical commodities as well as for specialty biochemicals such as amino acids and antibiotics, and even more complex products, such as vitamins, and enzymes.

This plot (Figure 1) shows several things. The slope of the line reflects the economy of scale in manufacturing. Products from the biochemical process industry fall on the same line as those of the chemical industry and take advantage of the same economies of scale experienced in commodity chemicals production.

In addition, the vertical distance from the abscissa reflects value added to the initial feedstock. Most biochemicals are made using inexpensive raw materials, such as sugar, and they offer good potential value added. The profit margin depends on the efficiency in transforming these raw materials into products. It is this biochemical problem that needs to be translated into a biochemical process. At this point one begins to see the need for integrating improved conversion yields, better metabolic pathways, and new reaction mechanisms. This requires integrating biochemistry, microbiology, and chemical process technology.

A number of things will stimulate success in biotechnical routes in manufacturing. One is higher prices for petrochemical feedstocks, which would make the use of biological routes to access renewable resources more important. Another is market growth for biological and chemical products that would require new manufacturing capacity. Continued development of genetic engineering is another stimulant, because this new technology permits doing things that could not be done before, as described below.

APPLICATIONS OF GENETIC ENGINEERING

Genetic engineering is an important tool for the chemical engineer as well as for the molecular biologist. From a chemical engineer's point of view, recombinant DNA and other techniques of genetic engineering permit several things that previously were not possible. First, the ability to introduce foreign DNA into new places means that microorganisms can be made to do or to produce something they did not do or produce before. Second, the expression of that DNA can be amplified into protein products. Third, biochemical pathways can be altered: for instance, an organism previously not capable of making compound C from compound B can be altered by the addition of a new enzyme in that cell to allow it to make compound C. Lastly, plant and animal metabolism can be altered through gene therapy. The discussion here will focus only on the first three capabilities.

The ability to introduce foreign DNA into a cell is especially important. From a process point of view it means several things. For instance, it is now possible to produce interferon, insulin, growth hormone, and other protein pharmaceutical products by fermentation. These are not new products, but microbial fermentation is a new way of making them at potentially much lower cost. There is also opportunity for making better products. When isolated from microorganisms rather than from natural sources, material can be produced that is more pure and that does not contain related proteins having other biological functions. It is also possible to use the technology of protein engineering. DNA contains the code for a unique sequence of amino acids that imparts the unique structure responsible for the catalytic or physiologic activity. By altering the DNA, hence altering the protein sequence and thus the structure and functionality of the resulting protein, it is possible to improve the final product. Several years ago this was a dream. Today it is reality, and several successful examples exist. Thus, one can perform molecular engineering to improve a product for the consumer or to improve a process. To achieve this goal it is important to understand the molecular basis for functions in order to manipulate the DNA structure. This understanding is still weak; nonetheless, protein engineering to make better products is very exciting technology.

It is possible to reduce manufacturing costs as well as to improve the end product. The ability to amplify DNA by causing a cell to make multiple copies of genes allows one to obtain more product per cell. This will lower the manufacturing cost because less energy, less labor, and less material are required for production. It is also possible to get higher productivity and higher purity. Thus, recombinant DNA is im-

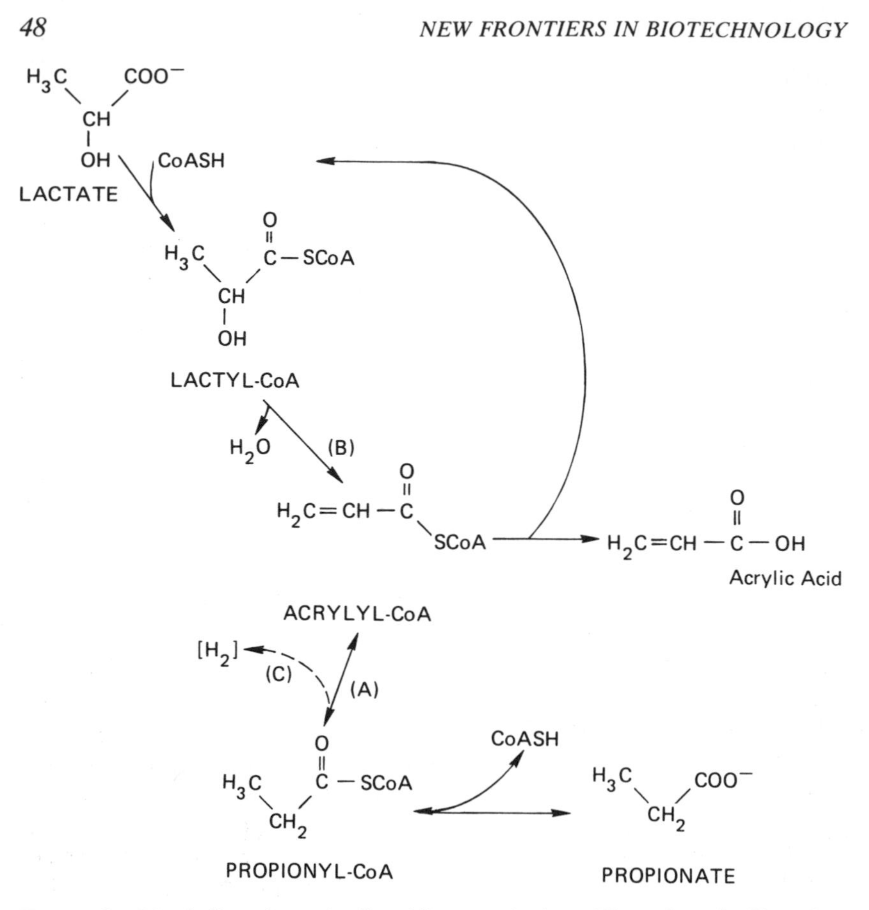

FIGURE 2 Metabolic pathways in *Clostridium propionicum*. The pathway in *Clostridium propionicum* for synthesis of acrylic acid is shown by the solid lines. If reaction A is blocked or eliminated, acrylic acid could be produced from lactic acid. If reaction B were blocked or eliminated and reaction C requiring a new enzyme (shown by the dotted line) were added, acrylic acid could be made from propionic acid.

portant to improving process technology, reducing manufacturing cost, and leading to new opportunities for process development.

The ability to alter biochemical pathways is important in terms of both old and new products—there are opportunities to develop better ways of making existing antibiotics or other biochemicals and to create new pathways for new products. The ability to alter biosynthetic pathways is important to manufacturing commodity chemicals that heretofore could not be made through biological routes. An example is acrylic acid, which is not a usual biochemical intermediate. However, using the pathway shown in Figure 2, it is possible to make cells excrete it and produce it by a variety of mechanisms. Through recombinant DNA one could conceivably improve this production route. In this way, through biotechnology, it is possible to gain access to new and cheaper raw materials for the production of commodity products.

By now it should be clear that genetic engineering can be used to address some of the problems of the CPI. It is important to understand this role of genetic engineering in both the existing biochemical process industry and in the chemical process industry because it has catalyzed much interest and investment in biotechnology. Let us now consider the realities of how one translates genetic engineering into a process that will deliver a product to the consumer, which is really the primary objective. In other words, the objective is not how to commercialize biotechnology, but how to use biotechnology to commercialize new processes and products.

BIOCHEMICAL PROCESS DEVELOPMENT

Figure 3 illustrates a typical biochemical process with its various unit operations. Raw materials are pretreated by heat to be sterilized and are then fed to the bioreactor. The bioreactor may be a traditional, batch fermentor or a novel device designed for a specific product. At this point in the process, value is added to the raw materials in synthesis of the final product. This is where genetically engineered organisms function. The important problems are how to control these reactors optimally; how to build them large enough to obtain economy of scale; how to get high productivity; what the limits in productivity are; and, once there is a product, which is invariably in dilute aqueous solution, how to recover it from the cell and the broth.

As a consequence of recombinant DNA, problems have arisen that

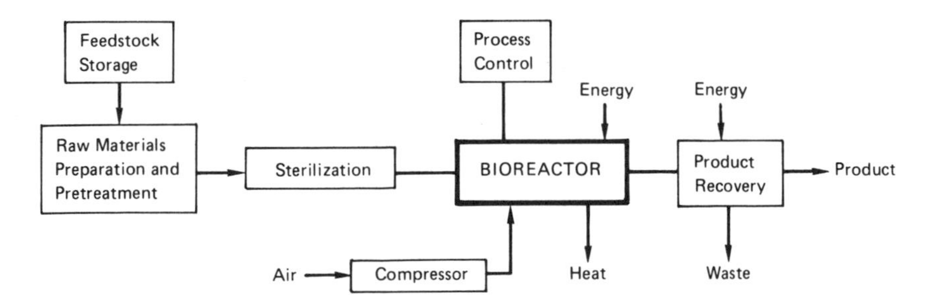

FIGURE 3 Schematic flow sheet for a typical biochemical process showing the major unit operations.

several years ago did not exist—for example, how does one recover a highly purified protein from a microorganism for use in the therapeutic marketplace? This was not a problem several years ago because pharmaceutical materials were not processed in this way. But today it is a problem, and the technology for protein recovery is not well established, especially for use in large-scale operations.

An examination of the biotechnology literature, both journal and patent, over the past 20 years reveals that 90 percent of the literature focuses on fermentation or bioconversion. Yet 90 percent of the problems and costs that limit the translation of scale from the test tube to the manufacturing plant exists downstream from the bioreactor and not in the bioreactor itself. While biocatalysis is enabling technology, recovery of product is needed for realizing a process.

What are the limits for biocatalytic processes? Some insight into this question can be gained from Figure 4, which shows productivity Q_p as

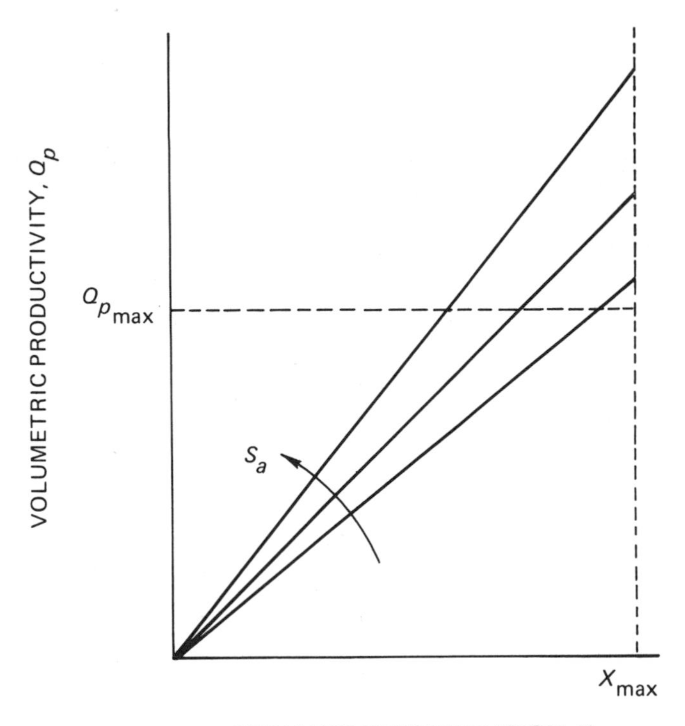

FIGURE 4 Volumetric productivity. Q_p (g/liter-h) as a function of catalyst concentration × (g/liter). S_a is the specific activity of the catalyst.

a function of catalyst concentration X. It is really the biocatalyst and its exquisite selectivity on which the potential of the biochemical process industry is based. There are two productivity limits. The first is that the maximum productivity is limited by transport phenomena, by the rate at which one can process biological materials. This is a fundamental heat and/or mass transfer limitation. Clearly, there are opportunities for the chemical engineer who has been dealing with these problems in chemical reactors for many years. This problem can also be considered as one of dealing with equipment limitations.

The second boundary is a limit on the amount of catalyst that can be put into the bioreactor. By increasing the specific activity, S_a, of the catalyst, which is the role of the molecular biologist and the microbiologist, productivity can be improved and cost can often be reduced. Improvement in productivity clearly requires cooperation between the engineer and the molecular biologist.

CONCLUSION

Finally, a series of problems in process development need to be solved in order to translate in scale the results of the molecular biologist to a process operating to deliver goods to the consumer. These bottlenecks in the development of biocatalytic processes are as follows:

- Limitations in productivity
  - Biocatalytic rate
  - Biocatalyst concentration
- High capital investment
- System stability
- New product routes
- Recovery processes

The problems include fundamentals of biocatalysis, heat and mass transfer, efficient conversion, and product recovery. Solving them requires integration in process development and integration between disciplines, and presents some very important challenges to those in engineering working closely with chemists and molecular biologists.

Biotechnology and Agricultural Research for Crop Improvement

CHARLES J. ARNTZEN

INTRODUCTION

Genetic improvement of crop plants has been underway for thousands of years. The first known organized plant cropping in the Iraqi Kurdistan (circa 6000 B.C.) involved seeding of wild wheat for cattle foraging. Since then our ancestors have harvested the largest and most desirable seeds or fruits of their favorite foods and planted some of these collections in organized agricultural efforts. During the last 50 years a better understanding of basic genetics and crop physiology has paved the way to dramatic increases in productivity of agricultural crop commodities via improved genetic and cultural practices (hybrid seeds, improved fertilizers, pesticides, and so forth). These approaches are being further refined at present in a wide range of academic and industrial settings, with continued, gradual increases in agricultural productivity.

A series of economic changes is currently affecting agricultural practices. Increased costs make it infeasible to expand irrigation, fertilizer, and other capital-intensive cultural inputs indefinitely. It is now recognized that it would be desirable to change the genetic makeup of our crops to give them increased resistance to environmental and biological stresses (e.g., heat, drought, nutrient starvation, insects, diseases) so that there will be less reliance on amelioration of these stresses by

Contribution number 11135 from the Michigan State University Agricultural Experiment Station. Research supported by a grant from CIBA-GEIGY and DOE Contract #DE-AC02-76ER01338 to Michigan State University.

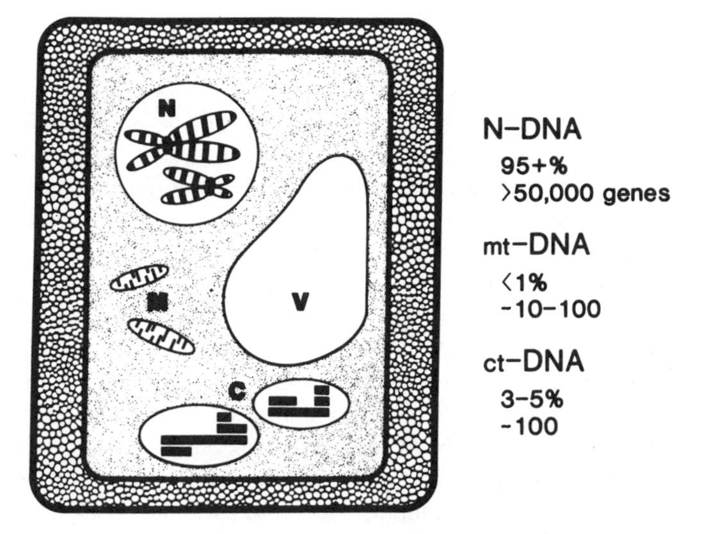

FIGURE 1 A diagram of the organization of a typical plant cell. The cell wall, surrounding the cell, is relatively rigid and defines the cell shape. Within the wall is the cytoplasm containing a vacuole (V), the nucleus (N), and several mitochondria (M) and chloroplasts (C). The structures diagramed within the nucleus are chromosones, the site of localization of 95 percent of the total DNA in the cell (corresponding to at least 50,000 individual genes). The rest of the DNA in the cell is in either the mitochondria (mt-DNA; about 10 to 100 genes) or the chloroplast (ct-DNA; also containing about 100 genes).

cultural means. Some of the desired traits were present to varying degrees in the progenitors of existing crop plants but have been lost in generations of crop evolution. In other cases the desired combinations of traits have never existed.

From the molecular biologist's standpoint, the current needs to rapidly add new physiological traits to our crops pose very interesting possibilities and problems. We must learn when we can use new biotechnologies to identify genes for desirable traits, and then put these traits into our present crops.

Location of DNA Within the Plant Cell

The genetic information of a plant resides in three different locations within the plant cell. Figure 1 shows in a stylized way what an individual leaf cell might look like in a crop plant. The cell has two major subdivisions. One is the cell wall surrounding the cell. The wall is largely inert, containing cellulose, lignin, pectin, and some protein. It forms a box or enclosure that gives shape to the cell (and to the plant as a composite of many cells). Within the cell wall is the cytoplasm, the living portion of the cell. It has various inclusions. In plants, as opposed to animals, a large portion of the cell is filled by a vacuole (analogous to a storage bag containing water, enzymes, salts, and organic molecules). Another major compartment is the nucleus, which contains several chromosomes. These contain about 95 percent of the DNA (deoxyribonucleic acid) in the cell. This is enough genetic material to code for an estimated 50,000 different genes.

It should be emphasized that it is not actually known how many genes are in the nucleus of crop plants or, in fact, what the vast majority of these genes do in controlling plant growth or development. It is known that the traits that are present have been selected through many generations of crop evolution. It is also known that most crop-breeding improvements have involved manipulation of this nuclear DNA by careful, time-consuming genetic crossing.

There are also two other sites where DNA is localized in plant cells. One is the mitochondria, the respiration centers of living cells (Figure 1). While plant mitochondria contain less than 1 percent of the total cellular DNA and approximately 100 structural genes, these cellular inclusions are still genetically important. Mitochondrial genes are known to influence disease resistance and male sterility, for instance.

The chloroplast is the third plant cellular compartment that contains DNA (Figure 1). These inclusion bodies also contain chlorophyll and are the site of photosynthesis (the conversion of solar energy to chemical form leading to the fixation of carbon dioxide [CO_2]). The organic molecules formed during photosynthesis are the materials upon which all life in our biosphere is dependent. In other words, plants are the only avenue for the input of energy in our biosphere. About 3 to 5 percent of the total cellular DNA is in the chloroplast, organized to include approximately 100 structural genes. It is known that these genes control the synthesis of important proteins that are involved in photosynthesis.

CHOOSING A TARGET FOR BIOENGINEERING EFFORTS IN CROP IMPROVEMENT

As was indicated earlier, plant cells contain tens of thousands of genes, each of which controls or influences a physiological process in the plant. At our present stage of research in molecular biology, we usually must deal only with one of these genes at a time. This limits the type of bioengineering efforts that can be contemplated in the near future.

1. Identify a (single gene) trait of agronomic value.
2. Determine the molecular basis of the trait (identify a critical enzyme, a rate-limiting step in a metabolic pathway, or the target site of a small molecule that regulates an important metabolic pathway).
3. Locate and isolate the gene that encodes the critical enzyme, or target site, and, where applicable, modify it.
4. Transfer the gene to a crop plant of choice.

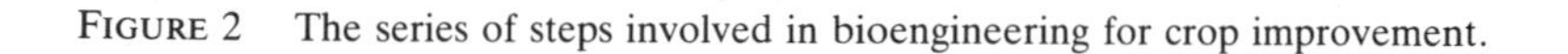

FIGURE 2 The series of steps involved in bioengineering for crop improvement.

Ultimately, crop yields are what everyone would like to be able to influence (increase) via biotechnology. Unfortunately, the yield of harvestable plant parts is determined by many traits (plant vigor, number of flowers, metabolic rates, and so forth). However, while there are many factors involved in the yield itself, we can try to limit our efforts to one specific item at a time.

The central question concerning biotechnology and agriculture now being asked in many academic and industrial laboratories is simple: How can we identify a feature of a crop (or crops) that we can change via genetic engineering to obtain a superior variety? Answering the question requires study of the progression of steps summarized in Figure 2.

At our current stage of bioengineering expertise, we must first find a single gene trait of agronomic importance and then identify the molecular basis for the trait. This could entail identifying a specific enzyme that produces a new or unique product. It could also mean identifying a rate-limiting step in some metabolic pathway which, if improved, would increase the rate of an important process. Another approach could entail finding the target site for a molecule that affects or regulates a specific metabolic pathway. Once we identify a key enzyme or target site, we can begin the process of using the tools of genetic engineering to change it. To accomplish this, of course, it is necessary to locate and isolate the gene that encodes the critical component. There are now a variety of strategies that have been applied to this problem, each too complicated to enumerate in any detail here, but still very feasible with today's technologies. It is also becoming a routine procedure to be able to make desired alterations in this gene, such as causing it to produce a protein with a different amino acid composition.

Once a gene is isolated, it is necessary to find a way to transfer the gene into the crop plant that we choose. This is an area of intensive research in many laboratories at the present time. Techniques (although limited as yet in applicability) are now available to achieve this gene insertion in certain crop plants. Advances in this aspect of biotechnology are needed and are certainly expected in the near future.

BIOENGINEERING FOR HERBICIDE RESISTANCE: ONE EXAMPLE OF GENETIC ENGINEERING STRATEGIES

Figure 2 delineated the general steps necessary for planning crop improvement via genetic engineering. The following material describes how this approach has been followed in one example.

The vast majority of major crops produced in this country are treated with pesticides. This provides a very cost-effective means of controlling weeds, insects, and other pests that would otherwise reduce productivity or quality of the commodity. In the production of corn, greater than 95 percent of U.S. acreage is treated with herbicides to control weeds. The most widely used herbicide in corn has been atrazine. The reason that atrazine can be applied to corn is that this crop has a very efficient metabolic pathway that detoxifies this and other triazine herbicides. In contrast, when most weed seeds germinate in the fields treated with atrazine, the herbicide is taken up and the weeds are killed.

The use of herbicides is not without its difficulties. One is carryover of the chemical in the soil. This results in problems when a farmer rotates crops so that a sensitive plant species is grown in a field that had previously been treated with a chemical that is toxic to the new plant. A common example is soybeans planted in rotation with corn. The triazine herbicides that are effective in corn can, if they carry over in the soil, severely stunt the soybeans grown on the same field in a subsequent year.

Normally, over the course of the growing season microorganisms in the soil degrade the triazines or other herbicides that are applied to the field. Under some conditions (such as a cold, wet spring), however, there is slow microbial activity. Problems arise in this situation, since the level of the residual herbicide can inhibit the crop to be planted.

The genetic engineer can see this circumstance as an opportunity for a contribution. If one could find a way to put triazine herbicide resistance in soybeans, the triazine carryover problem would no longer exist. The result would be of substantial economic benefit to the farmer.

Naturally Occurring Herbicide-Resistant Weeds

A second problem encountered in the use of herbicides is more recent. It centers around the appearance of new weed biotypes that are resistant to certain classes of herbicides. The most dramatic example is triazine herbicide resistance. This was first reported in 1970 but has now been documented for more than 30 species of weeds in various locations in the United States, Canada, Europe, and Israel. The problem is very analogous to insecticide resistance.

It is well known that houseflies became resistant to DDT within a few years after this insecticide's introduction. In the case of herbicides, resistance to the chemicals has taken many more years to develop. This is largely because of the much longer life cycle of plants compared with that of insects.

The new triazine-resistant weeds have frequently appeared in farms where corn has been grown in continual production with atrazine application year after year. It should be emphasized that the appearance of these weeds is currently more of an inconvenience to the farmer than it is a major problem. Weed control can still be achieved by switching to an alternative (although sometimes more costly) herbicide, since the resistance is specific for the triazine chemicals. The appearance of herbicide-resistant weeds must be taken as a portent of the future, however. If weeds become resistant to more and more chemicals, our options for weed control via herbicides will become more limited. In the meantime, however, the molecular geneticist can see a silver lining in this cloud—if we can establish how weeds became resistant to triazines, perhaps we can do the same for soybeans or other crops.

Research over the last five years has described in molecular detail the mechanism by which atrazine kills weeds. When the herbicide is taken up from the soil by germinating seedlings, it moves up from the roots to the leaves where it enters the chloroplasts.

The chloroplast (Figure 3) has a very simple but important function for the biosphere. It harvests sunlight. Chlorophyll is bound to the membranes inside this cellular inclusion. These pigments catalyze conversion of light into chemical energy, which is used to drive the carbon fixation pathways. The leaf takes CO_2 from the environment and converts it via photosynthesis into carbohydrates, amino acids, lipids, and so on. In this process the chloroplast also removes electrons from water, resulting in oxygen evolution, which is necessary for the respiratory activity of other living creatures in the biosphere.

The conversion of radiant energy (sunlight) into chemical intermediates involves many steps, each of which entails the movement of electrons along a chain of electron carriers (proteins with special functional cofactors) that are housed in the chloroplast membranes. When atrazine comes into the chloroplast, it binds to one of these electron carriers and blocks its function. This, of course, shuts off photosynthesis and thereby kills the plant by energy starvation.

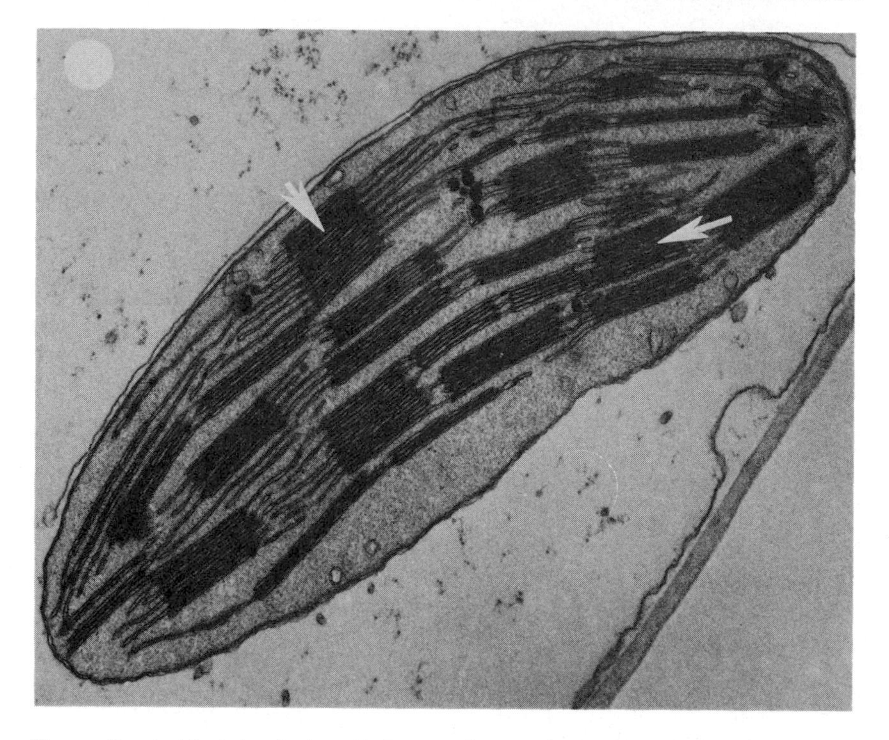

FIGURE 3 A chloroplast in the cell of a crop plant as observed in a thin-sectioned sample for electron microscopy. The oval-shaped chloroplast contains internal membranes (arrows) that are the sites of photosynthetic electron transport. One type of protein in these membranes binds atrazine and thereby blocks photosynthesis. As a result of mutation in the chloroplast DNA (the DNA molecules are too small to be visualized in this micrograph), a change in the herbicide-binding protein of some weed species has occurred. This makes the plant herbicide resistant. A research goal is to transfer the altered gene from the weeds to crop plants so the latter will also be herbicide resistant.

Molecular Basis of Triazine Herbicide Resistance

When atrazine-resistant weeds were discovered, several laboratories began investigating why the weeds did not die when this herbicide was applied. It was found that atrazine entered the new weed biotypes and was metabolized at low rates (unlike in the corn plants). Exclusion or detoxification mechanisms were therefore ruled out as explanations for why resistance occurred. Only when the chloroplast membranes were examined did an explanation become obvious—the atrazine could no longer bind to triazine-resistant chloroplast membranes. Obviously, it could not block photosynthesis and was therefore ineffective.

To learn why the atrazine did not bind to resistant chloroplasts, scientists first identified the protein that contained the herbicide binding site. It was then rapidly established that the gene for this protein resided on the chloroplast DNA. This led to the successful isolation of the gene encoding the herbicide-receptor protein. When the pair of genes from susceptible and resistant chloroplasts were analyzed to determine their nucleotide sequence, a single change was observed. This change causes the conversion of a serine in the protein of susceptible chloroplasts to a glycine in the protein of the resistant chloroplasts. We believe that this serine may play a central role in hydrogen bonding to atrazine and that it is thereby essential for the herbicide binding.

Genetic Exchange Between Weeds and Crops

Since the gene controlling the herbicide target site is altered in triazine-resistant weeds, the next question is: How can we make the same change in a crop plant? There is now one major success story along these lines. The work is the result of a novel breeding program conducted by university scientists at Guelph, Ontario.

The Guelph scientists had identified a new triazine-resistant biotype of the weed *Brassica campestris* (commonly called bird's rape). *Brassica campestris* has a large flower that is very amenable to genetic manipulation. It is relatively simple to transfer pollen (the male germ line) from one plant to the female flower parts of another plant. It should be pointed out that the pollen contains only nuclear genes and does not contain chloroplasts or chloroplast DNA. The egg cell of the female flower contains both the nucleus and chloroplasts. The chloroplasts, therefore, follow strict maternal inheritance in almost all crop plants. The Guelph scientists used a novel approach at this point. They took pollen from the flowers of oilseed rape (an important oil crop in Ontario and a close relative of the weed) and fertilized the egg of the weed flower. This means that the progeny of this cross will have half of the genes in their nucleus coming from the crop plant and half from the weed. The novelty of these experiments lies in the fact that all the chloroplasts came from the weed, since it was the female parent. This is a special case where we can document the genetic contribution of a wild relative to a crop plant's physiological properties.

The Guelph researchers continued this crossing, always taking pollen from the crop plant, for several generations. Since the genes of the weed were diluted out by 50 percent at each cross, they eventually ended up with a crop plant with cells containing the chloroplasts of the weed (which, of course, determine atrazine resistance). These scientists registered the triazine-resistant seed of oilseed rape in 1980. They have accomplished the same thing with summer turnip rape, another important crop in Canada. Again the trick was simple. They took pollen from the crop plant and used it to backcross onto the weed and the subsequent progeny of this cross until there was a nuclear substitution in the weed cytoplasm. In both cases just about all of the genes in the nucleus come from the crop plant, but the chloroplasts from the weed are left, and the result is herbicide resistance.

This novel crop-breeding program has opened up a new avenue for weed control in these two important crops. Unfortunately, this approach is not applicable to many other crop genetic systems, since there are no other cases where crops and triazine-resistant weeds are cross-fertile. This means that we must find other approaches, which is where new and novel bioengineering techniques are coming into play.

Fusion of Cells from Weeds and Crops

It is now routinely possible to isolate single cells from virtually all crop plants. When the cell walls are removed, the cells can be chemically induced to fuse. Current research in several laboratories is involved in fusing cells of *Solanum nigrum* (a common weed called black nightshade, which has developed herbicide resistance) and isolated cells of tobacco, potato, or tomato (crop plants that belong to the same family as that of *Solanum nigrum*). The object of these experiments is to donate chloroplasts from the weed to the cell line of the crop. In these experiments the nucleus of the weed cells can be inactivated chemically or by using X-irradiation resulting in new hybrid cells that contain the chloroplast of the weed plus the nucleus of the crop plant. There is one unconfirmed report that this strategy has been used to obtain a new tobacco plant that is triazine resistant. I anticipate the same success in the near future with potato and tomato and perhaps some other related crop plants.

GENE TRANSFER VIA MOLECULAR TECHNIQUES

The strategies for gene transfer outlined above (reciprocal crossing between cross-fertile weed and crop, and cell fusion to deliver weed chloroplast to crop plant cells) are straightforward. Unfortunately, these technologies are not now applicable to many major crops, including legumes (soybeans), cereals (wheat, rice), and other grasses. We must therefore project how these latter commodity crops can be manipulated. The most exciting approach will be to find a mechanism for transferring the gene of choice directly into a crop plant. In the past year great strides have been made in devising technologies for the incorporation of pieces of DNA (genes) into vectors (larger pieces of DNA, derived from a naturally occurring viruslike agent). These vectors can be induced to move into plant cells where the DNA they deliver is incorporated into the plant nucleus. Unfortunately, the atrazine-resistance gene described above resides in the chloroplast, so we cannot rely on existing technology to accomplish the gene transfer needed for the triazine-resistance problem. This only means a delay, however, since it seems certain that our understanding of the means of gene transfer into plants is only in its infancy. The triazine-resistance trait should, in fact, provide a tool with which biotechnologists can experiment to directly manipulate this and other genes.

CONCLUSION

It should be emphasized, in closing, that plant genetic engineering via biotechnology is a new and developing science that is going to supplement traditional agriculture. It will not replace standard methodologies for crop improvements, but it will add new facets. Traditional crop breeding will continue to be the mainstay of our agricultural system in production and release of new varieties. What biotechnology will offer is ways to increase the speed at which crop improvement is made, as well as providing tools for introducing novel traits that cannot be achieved with traditional genetics. It will be exciting during the next 20 years to see how new aspects of biotechnology will affect American and international agriculture.

Biotechnology for Health Care

J. PAUL BURNETT

Biotechnology has been an applied science in the pharmaceutical industry for a long time. The antibiotics that are used so extensively today in clinical medicine are products of fermentation or biotechnology. These substances have been produced on a very large scale for the last 30 or 40 years. Until a few years ago, however, the organisms used in biotechnology within the pharmaceutical industry were all isolated from nature. Existing organisms were selected using screening procedures designed to detect organisms producing useful substances.

Today the tools of biotechnology have changed. Molecular biologists have provided ways of designing and manipulating organisms to produce substances in which there is specific interest, rather than simply accepting what nature has already provided. This advance primarily accounts for the recent excitement within the pharmaceutical industry insofar as biotechnology is concerned.

In the pharmaceutical industry biotechnology generally encompasses two primary areas. The first is immunology, in which the discovery of hybridomas and cell-fusion technology have allowed the production of monoclonal antibodies. This has already led to some very important new diagnostic techniques, and it offers the promise of therapeutic applications in the future, but so far none of the latter has really been reduced to practice. This discussion focuses on the second area of biotechnology or biology that forms an important segment of the new technology in the pharmaceutical industry, namely, the area of recombinant DNA (deoxyribonucleic acid), where many discoveries have already been reduced to practice and where at least one product is already on the market.

The concept of recombinant DNA is based on the relatively well-understood function of DNA within all cells. A brief review of some basic concepts of molecular biology can help clarify the ways in which molecular biology and biotechnology can be used. Figure 1 shows the management information system of all cells. All cells contain genetic information stored in DNA. This information is transcribed from the DNA into a working pattern called messenger RNA (ribonucleic acid). This pattern is used in the cell to produce proteins. Proteins are the molecules in the cell that give rise to all of the characteristics that are recognized for particular cells.

Some of these protein molecules are enzymes, or biocatalysts, that catalyze the formation of all of the other molecules in the cell. Some

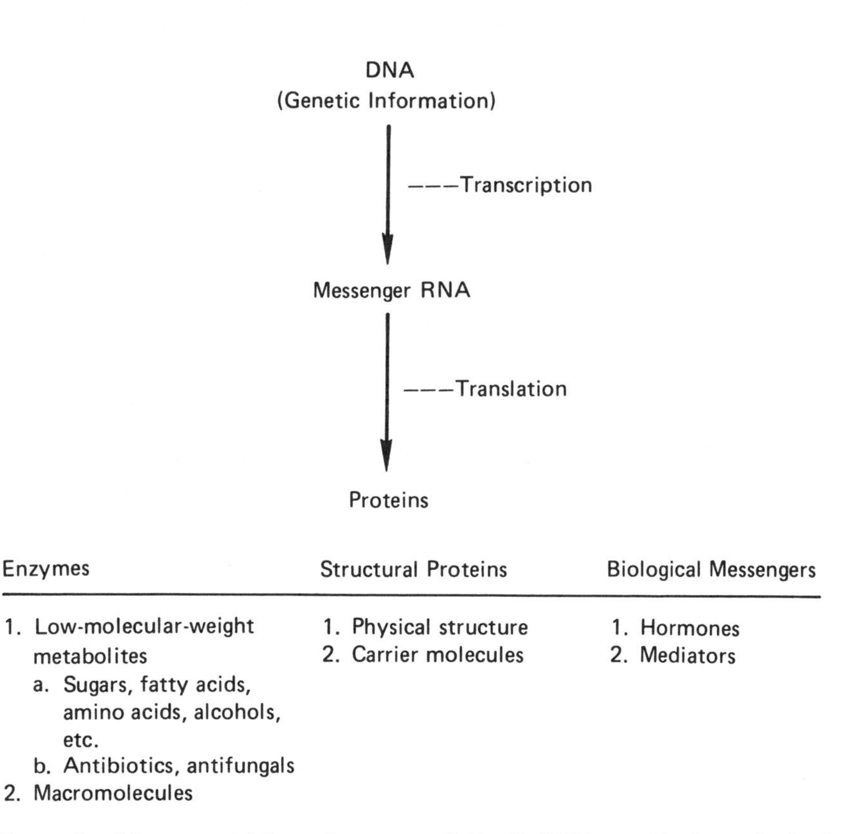

FIGURE 1 Management information system of all cells. DNA controls the synthesis of cellular proteins, which subsequently determine the phenotypic characteristics of the cell.

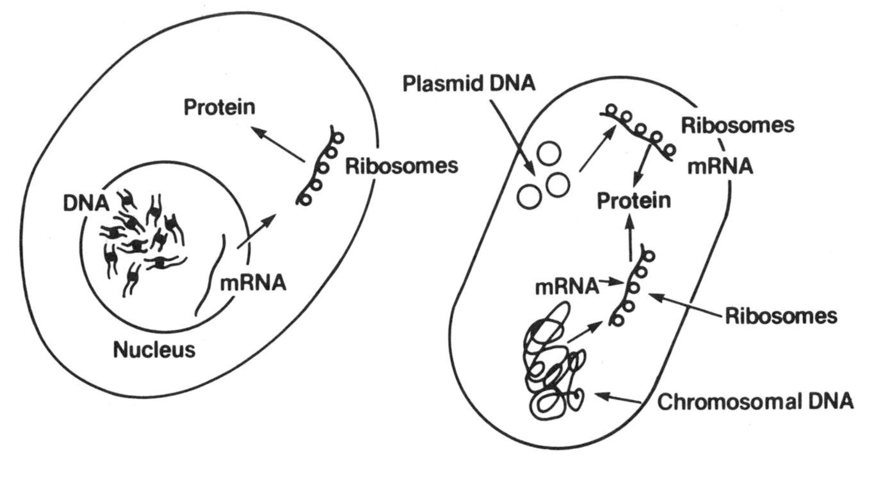

FIGURE 2 Pictorial presentation of management information of cells. In addition to chromosomal DNA, bacterial cells often contain small, autonomously replicating DNA molecules called plasmids.

of the proteins serve a structural function, giving rise to the physical appearance in the structure of the cell. Others serve messenger functions, carrying messages back and forth between the cells. The sum and substance of a cell is the complement of proteins that it contains.

Figure 2 presents much the same information as that in Figure 1, but in pictorial form, and it introduces an additional concept. On the left-hand side of the figure is a typical animal cell where DNA is in the form of chromosomes in the nucleus. The messenger RNA is made in the nucleus and goes out to the cytoplasm where it serves as a pattern for protein synthesis. The same process occurs in bacterial cells (right-hand side of Figure 2) except that bacterial cells have small DNA molecules (known as plasmids) that have particular advantages for use in biotechnology. As opposed to the chromosomal DNA, which is a very large and very fragile molecule, the plasmid DNAs are very small molecules. They can be isolated in a test tube. They can be cut apart and put back together. DNA can be added to them or subtracted from them. And, finally, plasmids can be put back into a cell in a functional form. Thus, plasmids form the cornerstone, if you will, of the application of molecular biology to biotechnology and recombinant DNA.

BIOTECHNOLOGY IN THE PHARMACEUTICAL INDUSTRY

Figure 3 indicates how, in a general way, one might use recombinant DNA to produce a substance via biotechnology in the pharmaceutical industry. In the upper left of the figure is a recombinant DNA organism being made by combining an animal gene with a plasmid DNA, followed by introduction into a microorganism. This first step, introducing the recombinant DNA into the microorganism, is a laboratory process. In the laboratory one would generate test-tube-scale cultures that contain the transformed cells that will now produce the protein coded by this animal gene. The next stages are development and production processes. This culture must be scaled up from the test-tube stage to the bioreactor, or fermenter, stage. The product must then be purified and packaged in suitable clinical form. Finally, before the product is ever subjected to clinical use, it is extensively tested in animal systems.

Is is important to point out that the bulk of this overall process begins

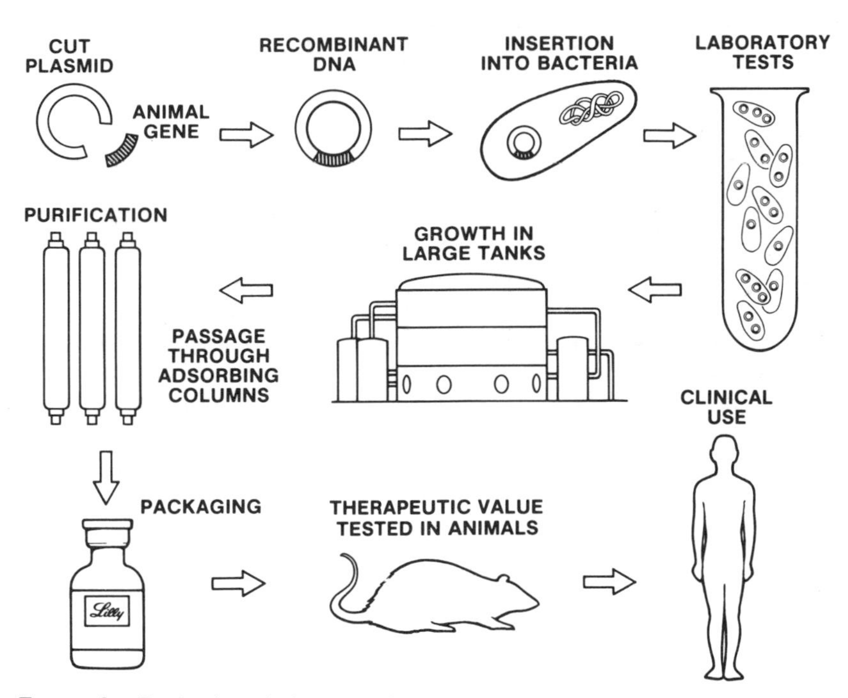

FIGURE 3 Production of pharmaceuticals by recombinant DNA. Recombinant DNA can be used to add new genes to microorganisms, and these can be grown in fermentation tanks to produce proteins on a large scale. Purification and extensive testing in animals precede clinical application in human beings.

TABLE 1 Amino Acid Residues and Molecular Weight of Human Polypeptides Potentially Attractive for Biosynthesis

Polypeptide	Amino Acid Residues	Molecular Weight
Insulin	51	5,734
Proinsulin	82	—
Growth hormone	191	22,005
Calcitonin	32	3,421
Glucagon	29	3,483
Corticotropin (ACTH)	39	4,567
Prolactin	198	—
Placental lactogen	192	—
Parathyroid hormone	84	9,562
Nerve growth factor	118	13,000
Epidermal growth factor	—	6,100
Insulinlike growth factors (IGF-1 and IGF-2)	70, 67	7,649, 7,471
Thymopoietin	49	—

after the genetic engineer has completed his or her work. In a sense, the contribution of the molecular biologist, although crucial, is a small portion of the total process.

Using the general process just described, a number of different types of molecules that have potential therapeutic use can now be made. The general categories of these substances include hormones and growth factors, pain-relieving proteins, plasma proteins, enzymes, proteins in the immunology area, and possibly even new types of antibiotics.

Table 1 lists some of the growth factors and hormones that one might consider producing by this technology. The genes for almost all of these proteins have now been cloned, and it is possible today to use those genes to produce these proteins in microorganisms. One at least, human insulin, has now been produced on a large scale and is a marketed product. It will be useful here to illustrate how genetic engineering is actually used to produce human insulin.

Production of Human Insulin

Theoretically there are two ways in which one could go about producing the insulin molecule (see Figure 4). Insulin consists of two different protein chains, the so-called A chain and the so-called B chain. One could produce the normal precursor of insulin, proinsulin, that is

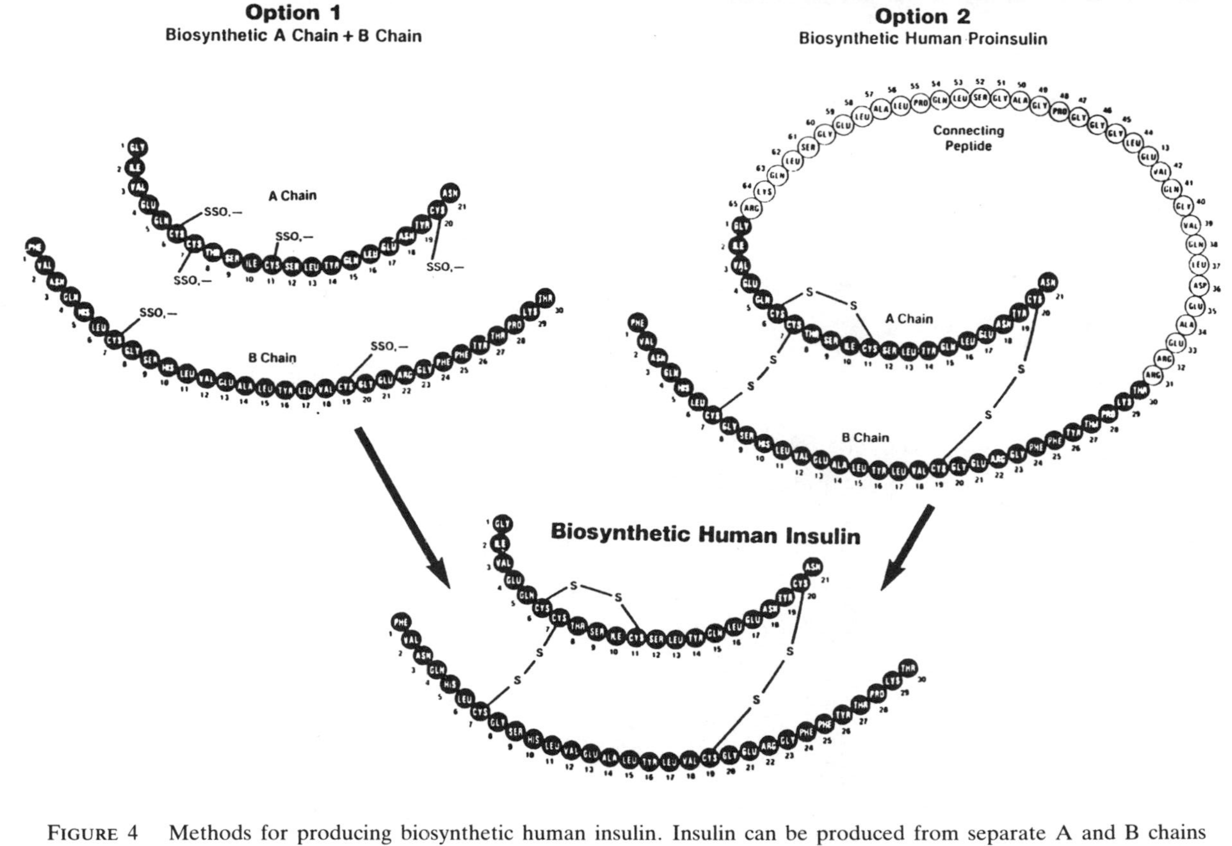

FIGURE 4 Methods for producing biosynthetic human insulin. Insulin can be produced from separate A and B chains synthesized individually in genetically engineered organisms. Alternatively, it can be derived from proinsulin produced as a single polypeptide chain in a genetically engineered organism.

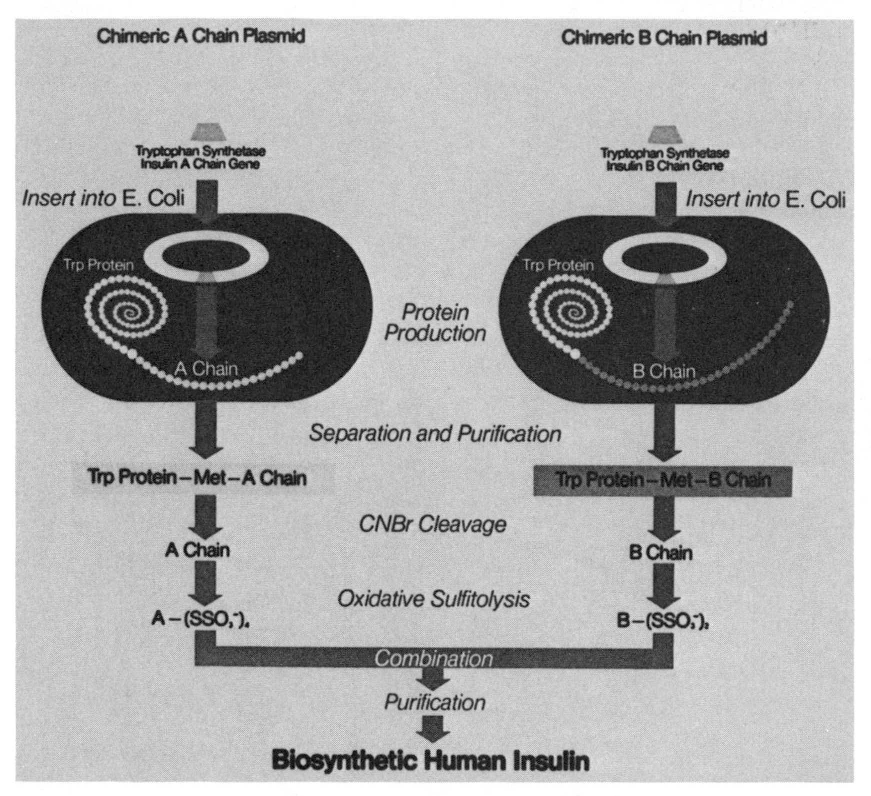

FIGURE 5 Schematic presentation of process for producing biosynthetic human insulin. Strains of *E. coli* have been engineered to produce insulin A and B chains. The initial cellular protein product is a chimeric protein in which the insulin polypeptide chain is attached at the carboxyl terminus of a protein coded by the tryptophan operon. Cleavage by cyanogen bromide releases the insulin chain from the chimera.

found in the pancreas. Proinsulin is a molecule that contains insulin but also contains an extra connecting peptide linking the two chains together. In the pancreas gland, this so-called connecting peptide is clipped out, leading to the production of insulin that is then released into the blood circulation system. One can mimic this process today in the laboratory and actually even in production, but up to the present it is not being used to produce human insulin on a large scale.

Presently the A chain and the B chain are made individually, and then these are coupled in the plant to produce the bioactive insulin. Figure 5 shows the process schematically.

In separate plasmids the genes have been introduced individually for the A chain and B chain of insulin, and then these plasmids have been transformed into bacterial cells. The *Escherichia coli* are then grown in large fermentation facilities. The product that is initially made is a large chimeric protein consisting of the A chain or B chain attached to the end of a naturally occurring *E. coli* protein.

This protein is subjected to a cleavage reaction in which the A chain and the B chain are chemically cleaved away from the rest of the chimeric molecule. Then, following several further purification steps, these two chains are combined, and the biosynthetic human insulin is recovered and purified.

Large amounts of the gene product of this plasmid accumulate in the *E. coli*. A thin-section electron micrograph of *E. coli* producing human insulin polypeptide shows dense areas, which are deposits of that protein within the cell. The protein is produced in very substantial amounts and can occupy a major portion of the cell.

This is one of the advantages of biotechnology today—by using appropriate control systems and regulatory systems on the plasmid being dealt with, one can make the protein of interest a major portion of the total protein of the microorganism. It can become a very efficient process.

Figure 6 shows crystals of the final product. It is a crystalline protein,

FIGURE 6 Crystals of biosynthetic human insulin produced by the process described in Figure 5.

and it has all of the characteristics of the insulin that is circulating in all of our bodies.

There are at least two advantages to being able to produce human insulin as opposed to continuing to use the pork and beef insulin that is currently used in many diabetic patients. First, the chemical structure of pork and beef insulin differs slightly from that of human insulin. Thus, there is the possibility of an improved therapy by using a molecule identical to the insulin that is already circulating in human bodies. The second advantage relates to the fact that currently produced pork and beef insulins are really by-products of the meat industry. Their production is subject to all of the economic pressures of the meat industry in terms of supply of pancreas glands. By production in microorganisms an essentially limitless supply of human insulin is available; the supply is no longer subject to the particular economic pressures of the beef and pork markets.

The following are some of the plasma proteins that one might consider producing by this technology:

- Albumin
- Globulins—α, β, γ
- Lipoproteins—α, β
- Plasminogen
- Fibrinogen
- Prothrombin
- Transferrin

Albumin, for instance, is a protein that can now be manufactured using recombinant-DNA technology. At least one company is working to scale this process up to commercial levels. Many of the genes for other proteins in the plasma protein series have also been cloned.

Other Uses of Recombinant DNA

Following is a brief discussion of examples of other ways in which recombinant DNA could be used to make products useful in the pharmaceutical industry.

Table 2 lists enzymes that are now used clinically. Probably the most important group today is that of the enzymes and cofactors involved in hemophilia—proteins like Factor VIIIc, which is used in hemophilia A. A large number of research groups are trying very hard to clone the gene for this protein; although the goal has not yet been reached, it can be expected that this will be accomplished in the near future.

One area of particular excitement is the fibrinolytic area. Blood-clotting problems, thrombosis, are an important aspect of clinical med-

TABLE 2 Ethical (Prescription) Enzymatic Products Currently Employed in the United States

Therapeutic Category	Enzymes Involved	Indications	Approximate Mfrs. Sales (millions of dollars)
Blood clotting factors			
Antihemophilic Factor (AHF)	Factor VIII	Hemophilia A	40
Plasma thromboplastin component	Factor IX	Hemophilia B	4
Gastrointestinal digestive	Pepsin, pancreatin	"Nervous" or other indigestion	11
	Pancreatin: lipase, trypsin, amylase	Inadequate fat digestion; cystic fibrosis	5
Wound debriding agents	Fibrinolysin and deoxyribonuclease	Removal of purulent exudates and eschar	4
	Trypsin		2
	Subtilains		1
	Collagenase		<1
	Streptokinase and streptodornase	Clot lysis, reduction of edema and inflammation	1
Oral proteolytic preparations	Chymotrypsin	"Possibly effective" for episiotomy	2
	Bromelains		
	Papain		3
Thrombolytic	Streptokinase	Lysis of intravascular blood clots	<2
	Urokinase		
	Fibrinolysin		
Absorption promoter	Hyaluronidase	Rarely for IM or SC inj.	<1
Ophthalmic surgery	Chymotrypsin	Cataract removal	<1
Cancer chemotherapy	Asparaginase	Leukemia	<1
Allergic drug reaction	Penicillinase	Destroy penicillin	<1

icine. A number of enzymes, such as streptokinase or urokinase, will degrade blood clots. The problem with both of these enzymes is that although they hydrolyze fibrin and destroy clots, they also lead to generalized bleeding in the patient. A new protein has recently been discovered, called the tissue activator of plasminogen (TPA), which is very specific and will hydrolyze fibrin only when it is in the form of a clot. This enzyme does not cause the side effect of generalized bleeding. The gene for TPA has been cloned, and one can now produce this protein using recombinant-DNA technology.

One of the most exciting areas for the application of biotechnology within the pharmaceutical industry is in immunology, where it will be

Factors Affecting Inflammation May Be Useful in	Helper Factors May Be Useful in	Suppressor Factors May Be Useful in
Patients with overwhelming infection	Tumor patients	Allergy
Postsurgical immune suppression	Aging	Autoimmune disease, SLE, arthritis
Burn patients	Diabetics	M.S., thyroiditis, myasthenia gravis, etc.
Postsurgical peritonitis	Dialysis patients	Transplantation
Tumor patients	Immunodeficiency disease	
Aging	Allergy	
Immunodeficiency disease	NK cell activity	
Dialysis patients	Burn patients	
Thyroiditis	Trauma	
M.S.-EAE	Postsurgical immune suppression	
Allergy	Chronic diseases—hepatitis, parasitic Hodgkin's disease	

FIGURE 7 Clinical applications of cytokines.

possible to produce some interesting vaccines. Instead of discussing these possibilities, however, let us turn to a group of proteins called cytokines. A cytokine is a molecule made in a cell as a result of some sort of stimulus. It can result from various types of stimuli. A cytokine is elaborated by the cell producing it. It leaves that cell and acts as a messenger, one of the functions of proteins mentioned earlier. It then stimulates a second type of cell to cause some sort of biological effect. It may affect the growth of the cell, it may affect the movement of the cell, or it may activate the cell to perform a specific function. For instance, in the phagocytic series it may activate a cell to become more active phagocytically. Figure 7 illustrates some of the areas where cytokines might be used clinically. Many cytokines are involved in the inflammatory response. Others promote the immune response, and still others are known to suppress the immune response. It is thought that the clinical states shown in Figure 7 represent those in which many of these cytokines might be used therapeutically. The important point with regard to this figure is that it represents a very large number of clinical diseases.

Figure 8 shows some of the cytokines that are produced by one particular type of human cell, the lymphocyte. It can be seen, for instance,

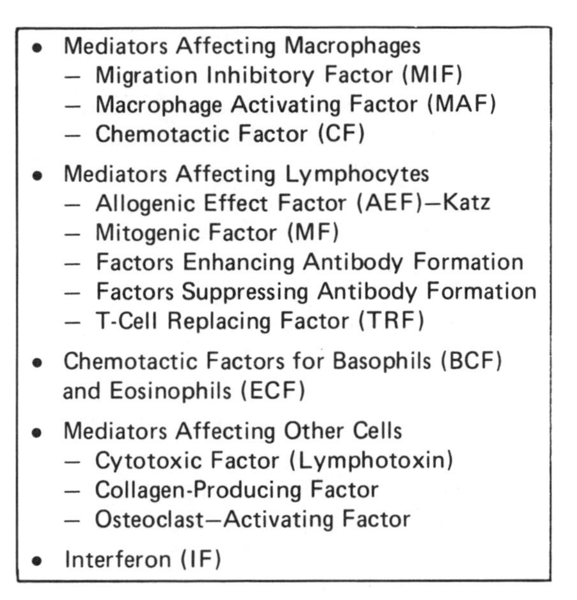

FIGURE 8 Cytokines produced by lymphocytes.

that there are mediators made that affect macrophages, the cells that destroy invading organisms in the body. There are mediators that affect other lymphocytes and cause them to do a variety of things. Some may enhance, for instance, antibody formation. There are chemotactic factors elaborated by the lymphocyte that affect cell movement. One of the cytokines that has been most publicized in the recent past is interferon, which is made by the lymphocyte, among other types of cells.

All of the examples just examined illustrate the type of products that can reasonably be expected to be produced by biotechnology in the pharmaceutical industry. However, exciting as these present applications for recombinant DNA may be, it appears that the total number of drugs that will be produced by this technology as proteins will be limited. In the future the greatest value of the application of biotechnology and recombinant DNA within the pharmaceutical industry will probably come about as we begin to understand life at the molecular level.

In addition to offering a way of producing molecules, recombinant DNA offers the molecular biologist a way of cloning and isolating genes, characterizing these genes, and understanding their function at a genetic level. In fact, today it is sometimes much easier to isolate the gene for a particular protein and to learn the structure of the protein from the gene than it is to isolate the protein itself.

All of the processes of life, and especially those orderly processes such as differentiation and development seen in the normal growth of plants and animals, are ultimately controlled by DNA. This is also true of the disorderly processes that we recognize, for instance, malignant cell growth. As we define these life processes at a genetic level, it is expected that it will be possible to design new small molecules that may be produced by traditional chemical means which will be the drugs of the future; but the discovery of these drugs will hinge on the application of molecular biology.

Following is one example of how that might happen and of one area where this might be applied. Genes known as oncogenes have been discovered; they have been found in a variety of tumor viruses from various types of animals. It is also known that there are corresponding genes in normal cells that seem to be very similar in structure to the oncogene of the tumor virus. When a tumor virus invades a cell, the function of the oncogene is to produce a protein that leads then to the transformed or malignant phenotype. The function of the corresponding gene in the normal cell is to regulate the growth and function of the cell.

We believe that it will be possible in the future, by understanding how these genes are genetically regulated and how these proteins function, to develop small molecular inhibitors that may be intriguing new types of chemotherapy for treating cancer. It is certainly too early to say what form these will take or exactly what structure they might be, but it is an interesting example at least of a way in which recombinant DNA might lead indirectly to new drugs.

CHALLENGES TO BE MET

Today we can use DNA technology to produce drugs that have important clinical uses, and in the future we can expect products, as yet unimagined, that will flow from our increased understanding of the biological processes brought about by the use of recombinant-DNA technology as a research tool. However, further problems must be solved in order for there to be broad application of biotechnology to the production of proteins in the pharmaceutical industry. These challenging problems involve the following:

- Fermentation
    - Regulation of protein production
    - New host organisms, including mammalian cells
    - Fermentation technology development
- Protein recovery and purification

Many of the problems faced today in the application of recombinant DNA within the industry relate not to molecular biology and to how one genetically engineers a cell, but to how the production of protein in the fermentation vessel itself is regulated once that cell has been genetically engineered.

We know today that there are proteins that probably will not be able to be produced economically in bacterial cells. We believe that there will be occasions in the future where mammalian cells will have to be used to produce those proteins. So the biotechnologists or bioengineers in the fermentation industry will have to learn to deal with mammalian cells on a large scale or in mass culture.

It is my belief that the companies that develop their fermentation technology in general to the greatest degree will be the most successful in applying recombinant DNA and biotechnology.

There will be need for great improvement in protein recovery and purification. Many of the techniques used today on production scale, for instance, to produce human insulin, really mimic the techniques of the laboratory. There is great opportunity for new innovation particularly in the area of protein recovery and purification.

And finally, it appears that the organizations that will most successfully apply biotechnology in the future will be the companies that can bring about the closest collaboration between the genetic engineer and the biochemical, electrical, and other engineering disciplines.

Part III

Advances in Structural Materials

Introduction

ALBERT R. C. WESTWOOD

The dramatic and well-publicized advances in computers and biotechnology of recent years have overshadowed those in structural materials. But the materials community has not been resting on its laurels. Instead, during the past decade it has generated a variety of mechanistic insights and innovative ideas, some of which are now beginning to emerge in engineering applications.

For example, although the potential usefulness of composite materials has been evident for some time, it was necessary to learn how to produce complex shapes from them and how to join these together before engineering structures could be produced. These steps have now been accomplished. To illustrate: welding silicon-carbon (SiC)-reinforced aluminum initially proved to be difficult, because hydrogen trapped in the metal concentrated into voids in the weld zone producing a weakened, Swiss-cheese-like structure. But treatments to remove this hydrogen prior to welding have been devised, and now, stiff, lightweight, metal-matrix composite structures, such as the yoke for a shipboard satellite communications antenna,[1] are beginning to be produced.

Fiber-reinforced plastics (FRP) also are emerging as major structural materials. The new Lear Fan Turboprop, for example, makes extensive use of carbon-fiber-reinforced epoxy to save 40 percent of the weight of a similar structure in aluminum. Without engines this airplane weighs only 1,275 pounds. With engines it can travel 2,300 miles at up to 400 miles per hour on 250 gallons of fuel.[2] Industry sources project that up to 65 percent of the structures of commercial aircraft will be made from composite materials by the mid-1990s.

In the case of the Lear Turboprop, FRPs serve both aesthetic and structural purposes, but these functions are separated in the new Pontiac Fiero automobile. Glass-flake-reinforced, injection-molded polyurethane is used for the vertical surfaces of the Fiero, and this can bend and snap back on minor impact, reducing body damage. The easily removed plastic panels simplify body repair and reduce corrosion.[3]

Progress has also been made with inorganic materials. Portland cement has been the most extensively used and cheapest material of construction for more than 100 years, but unless it is reinforced with steel it can be used only in compression because of its poor tensile properties, a consequence of the presence of pores that act as crack nuclei.[4] However, by controlling ingredient particle size, adding surface-active agents to improve the rheology of the paste and polymers to fill in the pores, and working the mixture to remove entrapped gas bubbles, a Macro-Defect-Free (MDF) cement can be produced, the bend strength of which approximates that of aluminum and from which springs can be fabricated (Figure 1).[5] Future developments along these lines could markedly influence both architecture and the durability of roads.

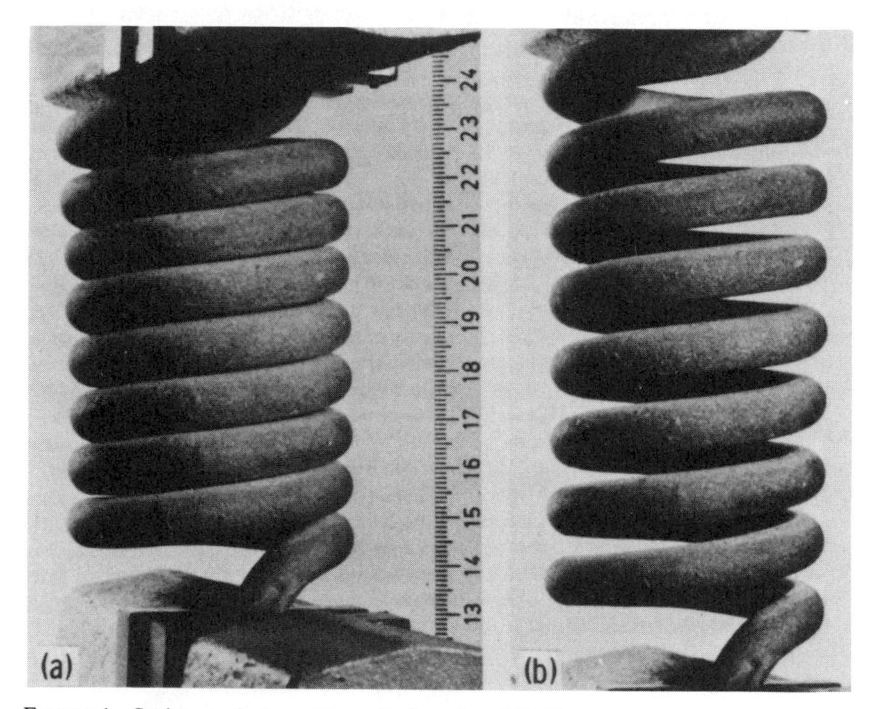

FIGURE 1 Spring made from Macro-Defect-Free (MDF) cement in (a) relaxed and (b) extended positions.[5] Reprinted with permission.

FIGURE 2 Structure of tough, strong abalone shell, consisting of platelets of $CaCO_3$, ~0.2-μm-thick, interleaved with layers of protein. From Note 4. Reprinted with permission.

The notion that ceramics can be tough is intriguing, since ceramics are usually thought of in terms of rather fragile pottery. However, by incorporating crack arrestors (e.g., fibers) and crack energy dissipators (e.g., phase-transforming particles) into the structure, ceramics and ceramic composite materials exhibiting quite useful measures of toughness (20 megapascals square-root meter [$MPam^{1/2}$]) can now be produced. Early applications for such materials include scissors (such as those made from partially stabilized zirconia by Toray Industries of Japan) and surgical saws.

Of course, tough inorganic structures are really nothing new. Figure 2 illustrates a section through an abalone shell. It consists of 0.2-micrometer (μm)-thick platelets of 99 percent pure $CaCO_3$, interleaved with thin layers of protein. Its tensile strength also is about that for a conventional aluminum alloy, and it is as tough as Plexiglas.[4]

Inorganic glasses have been with us for millennia, but metallic glasses are a relatively recent arrival, and they represent one of an increasing group of materials whose useful and different properties are a consequence of the application of very high rates of cooling from the molten state.[6] Amorphous iron-base alloys can provide magnetic losses 10 times smaller than those of the traditional Fe-3% Si materials, and half those of the permalloys. Amorphous metals also exhibit remarkable strengths and resistance to corrosion, the latter property being attributed to the absence of second phases and grain boundaries.

However, of currently more practical value as structural materials are the rapidly solidified (RS), crystalline alloys of aluminum (Al), nickel (Ni), and cobalt (Co). Produced using quenching rates of order 10^{6}°C per second or more, they are characterized by a very fine grain size, reduced amounts of segregation, and extended solubility of alloying elements.

RS aluminum alloy components are now entering service, e.g., on Boeing 757 aircraft, providing improved strength-to-weight ratios and

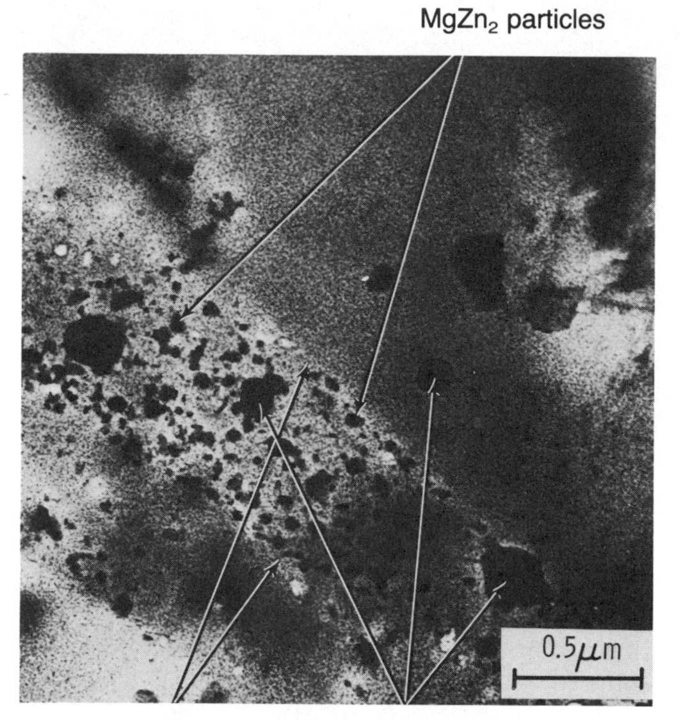

FIGURE 3 Structure of rapidly solidified cobalt-containing aluminum alloy. Co_2Al_9 particles are small in size and well distributed throughout matrix and grain boundary.[7]

resistance to stress corrosion. To illustrate, cobalt has very limited solubility in aluminum, and concentrations of only 0.5 percent or so in ingots produced by conventional metallurgical processes result in the precipitation of large cobalt intermetallic phases that reduce fracture toughness. However, when aluminum alloy powders (7000 series) containing 0.4 to 0.8 percent of cobalt are prepared by rapid solidification and then consolidated by compaction and extrusion, the Co_2Al_9 intermetallic compound appears as fine particles dispersed throughout the structure (Figure 3), modestly enhancing strength, but markedly increasing resistance to stress corrosion cracking (SCC). In fact, the critical stress for the onset of SCC is increased by about 25 percent in the 0.4 percent Co alloy and 50 percent in the 0.8 percent Co alloy. This improvement is thought to be caused in part by the Co_2Al_9 particles acting as hydrogen recombination catalysts, reducing the likelihood of hydrogen atoms entering the aluminum alloy and causing embrittlement.[7]

Other developments include treatments to change the chemical composition or structure of the surface of a solid to improve its mechanical properties or durability. With wear and corrosion costing U.S. industry an estimated $80 billion to $90 billion annually, this is clearly a research area of economic significance.

One approach involves improving near-surface properties without changing the composition of the material. This can be achieved, for example, by the intense local heating of a laser beam. Figure 4 shows a carbon dioxide (CO_2) laser being used to harden the wear surface of a crankshaft.[8] A second approach involves changing the near-surface

FIGURE 4 Co_2 laser beam being used to harden lobe surface of crankshaft. From Note 8. Reprinted with permission.

chemistry or microstructure of the solid, for example, by means of ion implantation. Using this approach the surface of an iron component can be implanted with chromium and made as corrosion resistant as that of stainless steel. Likewise, it has been found that the wear resistance of a diesel fuel injection pump can be increased a hundredfold via yttrium implantation.[8]

Chemical treatments that alter the structure of oxide films also are proving useful. Such films can be made porous and then filled with Teflon, graphite, or MoS_2 to produce hard, corrosion-resistant, and lubricious surfaces. The morphology of oxide surface films also deter-

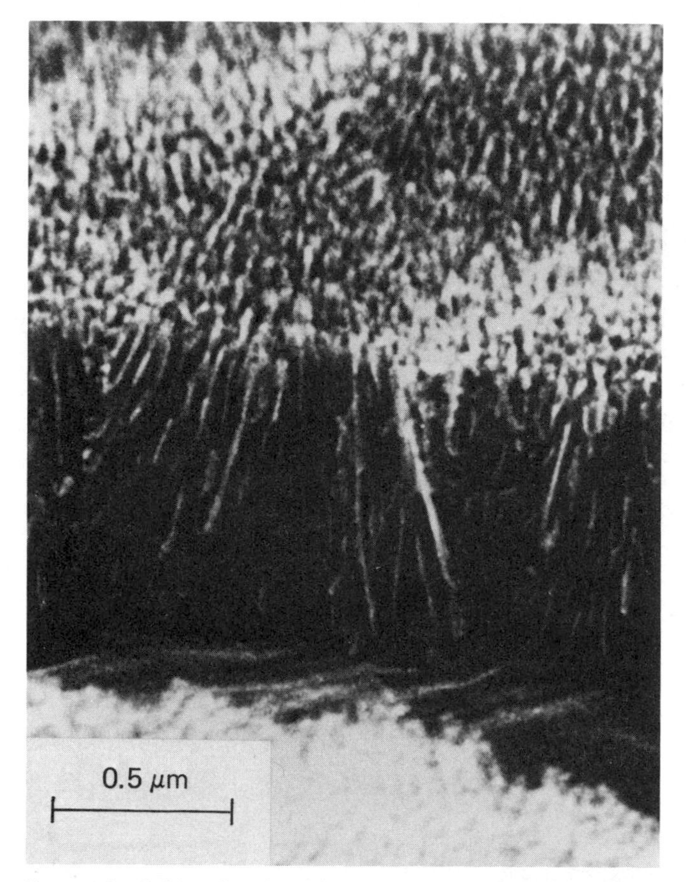

FIGURE 5 Oxide cell and whisker structure produced on aluminum by an acid pretreatment. Such structures permit the development of strong and reliable adhesive bonds in aircraft manufacture.[9] Reprinted with permission.

mines the strength of structures made by adhesive bonding, as aerospace structures increasingly are. A smooth oxide film on aluminum produces weak and unreliable bonds, but an acid pretreatment that produces a porous oxide film with a cell and whisker structure, such as that shown in Figure 5, to which the adhesive can become mechanically as well as chemically attached, produces a strong and reliable bondment.[9] The durability of adhesively bonded aluminum structures can be further improved by first spraying the oxide film with a dilute aqueous solution of NTMP (nitrilo-tris-methylene phosphonic acid). The adsorbed molecules of this and similar organic substances inhibit the transformation of the strong oxide whiskers in humid, salt-containing environments into weak, stress-raising, hydroxide platelets.[10] The result is a severalfold increase in bond life.

In summary, new types of stronger, more durable, and increasingly cost-effective structures are beginning to emerge, with the intrinsic properties of all classes of solids—metals, ceramics, polymers, and composites—being more efficiently utilized. Indeed, as R. B. Nicholson[11] recently commented, the future role of materials science lies not so much in developing new materials per se as in developing new and more efficient ways of processing existing materials so that they exhibit the properties of which they are theoretically capable.

Such efforts will provide us with strong and tough ceramics, extremely corrosion-resistant metals with outstanding strength-to-weight ratios, more durable concretes for road use, electrically conductive and self-reinforced polymers, and "designer" composites with a wide variety of combinations of desirable properties.

NOTES

1. *Materials Highlights*, Naval Surface Weapons Center, White Oak, Md., Feb. 1982.
2. *Mater. Eng.*, p. 43, May 1982.
3. *Mater. Eng.*, p. 30, Oct. 1983.
4. J.D. Birchall and A. Kelly, *Sci. Am.*, p. 104, May 1983.
5. J.D. Birchall, A.J. Howard, and K. Kendall, *J. Mater. Sci. Lett.*, *1*:125, 1982.
6. *Amorphous and Metastable Microcrystalline Rapidly Solidified Alloys*, National Materials Advisory Board Rep. No. 358, National Research Council, Washington, D.C., May 1980.
7. J.R. Pickens and L. Christodolou, unpublished work, Martin Marietta Laboratories, 1983.
8. C. Rain, *High Tech.*, p. 59, March 1983.
9. J.D. Venables et al., *Appl. Surf. Sci.*, *3*:88, 1979.
10. J.S. Ahearn et al., in *Adhesion Aspects of Polymeric Coatings*, Plenum, New York, 1983, p. 281.
11. R.B. Nicholson. Reported in *Chem. Eng.*, p. 36, March 1983.

Rapid Solidification Technology

BERNARD H. KEAR

Ever since the pioneering work on "splat quenching," reported by Duwez et al. in 1960,[1,2] it has been known that rapid quenching from the molten state, i.e., rapid solidification, is a means to develop unusual, even novel microstructures, which frequently exhibit beneficial properties. In order to exploit such structure/property advantages, much effort has been expended, at least since about the mid-1970s, on developing new methods for (1) production and consolidation of rapidly solidified fine powders, (2) fabrication and utilization of rapidly solidified thin filaments or ribbons, and (3) rapid solidification surface treatments of materials. This paper will highlight some of the more exciting innovations that have occurred in these areas. As will be shown, major advances have been made on all three fronts, with real prospects for the widespread use of rapidly solidified materials in structural and magnetic applications.

Considerable progress has also been made in our understanding of rapid solidification behavior, including mechanisms and kinetics of rapid solidification, phase transformations, and structure/property/processing relationships. Although a complete discussion of such fundamental aspects is clearly beyond the scope of this paper, some pertinent findings with respect to the influence of cooling rate on solidification microstructure and the effects of subsequent heat treatment will be briefly examined.

Only selected aspects of rapid solidification technology are discussed here. For more complete information on the subject, the reader is referred to recent publications.[3-6]

MICROSTRUCTURAL CONSEQUENCES OF RAPID SOLIDIFICATION

The principal effects of rapid cooling from the molten state on the resulting solidification microstructure are summarized in Figure 1. Under ordinary casting conditions, with cooling rates of ~1 kelvin per second (K/s), the microstructure typically is very coarse and exhibits a high degree of chemical segregation. In passing from ordinary casting practice to cooling rates >10^2 K/s, there is a progressive refinement in the solidification microstructure, i.e., dendrites, eutectics, and other microconstituents are all reduced in scale. This is because with increasing cooling rate there is much less time available for coarsening of the microstructure. The degree of segregation within these structures, however, remains essentially the same, since local equilibrium is maintained at the solid/liquid interface during solidification, such that local temperatures and concentrations are given essentially by the equilibrium phase diagram. In other words, the indicated microstructural refinement is a consequence of differences in the growth process rather than of effects due to undercooling of the melt prior to the nucleation stage.

With increasing cooling rates substantial undercooling of the melt can occur, and it is in this solidification regime that novel microstructures make their appearance. These are indicated in Figure 1 as extended solid solutions, metastable crystalline phases, and amorphous metallic solids. Large departures from local equilibrium at the solid/liquid interface can occur in this solidification regime, with the solid phase entrapping supersaturated concentrations of solute and impurity atoms. In the limit, at sufficiently high cooling rates, the resulting solid will have exactly the same composition as that of the parent liquid. This

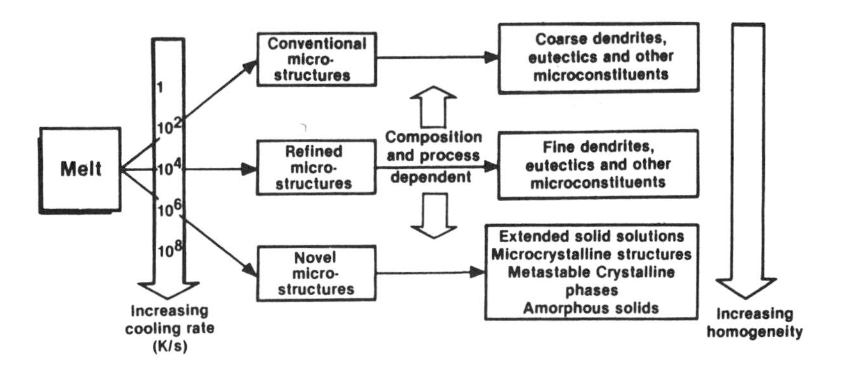

FIGURE 1 Microstructural consequences of rapid solidification.[3] Reprinted with permission.

mode of solidification is called partitionless, segregationless, or massive solidification, and its characteristic feature is the formation of an ideally homogeneous structure.

An additional complication at sufficiently high undercooling is the formation of a microcrystalline structure, due to the combined effects of higher nucleation rates and lower growth rates at the lower temperatures. In general, with increasing cooling rate, conventional alloys follow this sequence: coarse dendrites → fine dendrites → homogeneous or extended solid solutions → microcrystalline solid solutions. On the other hand, alloys that exhibit deep eutectic troughs tend to follow this sequence: coarse eutectic → fine eutectic → ultrafine eutectic → amorphous metallic solid.

It should be emphasized that the picture of solidification behavior depicted in Figure 1 is very approximate. The more correct picture must take into account the operative temperature gradient in the liquid phase just ahead of the advancing solid/liquid interface, and the interplay between temperature gradient and solidification rate, or interface velocity. Steep temperature gradients tend to stabilize plane front growth, with compositional homogeneity, whereas steep solute gradients promote cellular or dendritic growth. These effects can be negated at sufficiently high interface velocity, where plane front solidification can occur irrespective of the operative temperature gradient.

An example of the refinement in dendritic structure observed in aluminum (Al) alloys with increasing cooling rate is shown in Figure 2.

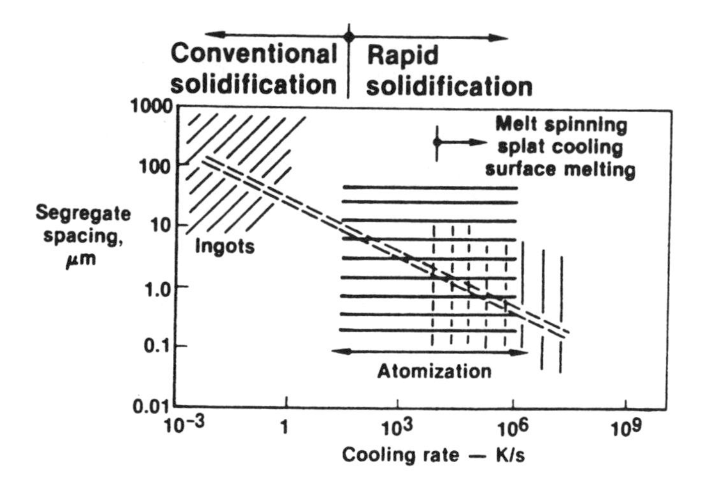

FIGURE 2 Effect of cooling rate on segregate spacing in aluminum alloys (dendritic mode of solidification).[3] Reprinted with permission.

The indicated segregate spacings were derived from measurements of secondary dendritic arm spacings. The reduction in segregate spacing by two orders of magnitude due to rapid solidification is of particular significance from the viewpoint of achieving compositional homogeneity by subsequent heat treatment. Thus, typically, annealing times are reduced from hours to seconds in many alloy systems where dendritic growth cannot be avoided even under the highest available cooling rates.

Experience has shown that the attainment of an ideally homogeneous structure, irrespective of whether it is accomplished by massive solidification or by heat treatment of refined dendritic structures, imparts real property benefits to the alloy. For example, in nickel (Ni)-base superalloys, which are prone to dendritic segregation, until the advent of rapid solidification processing the full benefit of γ^1 precipitation hardening was never achieved. It is now known that the optimum properties in such alloys can be realized only when the γ^1 precipitation hardening phase is uniformly distributed, which is possible only in homogenized material. Similar considerations apply to other alloy systems. In particular, it may be noted that the effective dispersal of extraneous phases in many alloys due to rapid solidification can give rise to unexpected benefits. For example, the fine scale dispersal of manganese sulfides in steels prevents grain coarsening during austenitizing treatments.[3] This result has also raised interesting questions concerning the possibilities for deliberately exploiting fine dispersions of sulfide phases in steels for hardening purposes.

RAPIDLY SOLIDIFIED POWDERS

Powder Production

Inert gas atomization[7] and centrifugal atomization[8] (Figure 3) are the most widely used methods for producing bulk quantities of rapidly solidified powders. Inert gas atomization involves the interaction of a melt stream with a symmetrical arrangement of converging high velocity gas jets. Atomization occurs as a result of the dissipation of gas phase kinetic energy in the interaction zone. Most commonly the working fluid is steam, nitrogen, or argon. Centrifugal atomization employs a high speed rotating disc atomizer for particle generation and high mass flow helium (He) gas for quenching purposes. Good wetting between melt and disc surface is a prerequisite for efficient powder production. This is achieved by forming and maintaining a thin solid skull on the surface of the water-cooled copper disc. The forced convective cooling employed in centrifugal atomization generates high cooling rates, typically ~10^5 K/s for

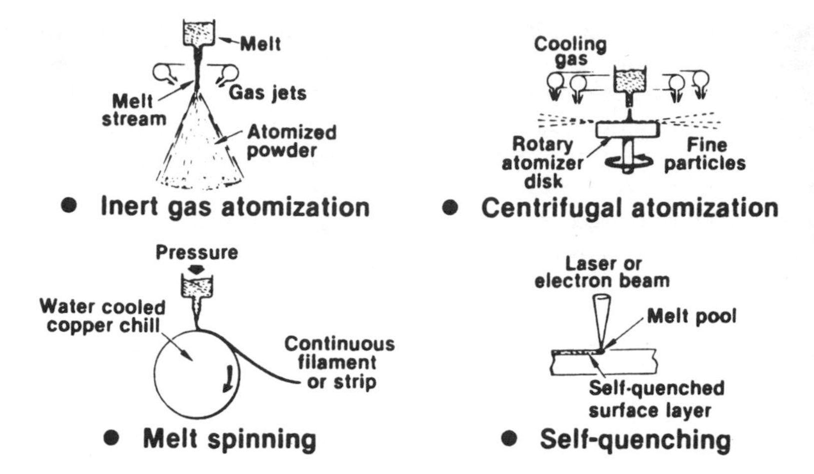

FIGURE 3 Representative rapid solidification processes.[3] Reprinted with permission.

particles ~50 micrometers (μm) in diameter (dia.). The cooling rate for a comparable particle size in inert gas atomization is ~10^4 K/s.

Both processes yield spherical particles (Figure 4) in a size range of 20 to 100 μm dia. In conventional alloys the microstructure is typically refined dendritic. In certain alloys, amorphous structures can also be developed but only in smaller particles (<10 μm dia.) that experience the highest cooling rates (~10^6 K/s). A more convenient method for fabricating high yields of amorphous powders is by pulverization of amorphous melt-spun ribbons. The resulting powders exhibit clean, smooth fracture surfaces and have a gritlike appearance (Figure 4).

Conventional Consolidation

When the principal benefit of rapid solidification is perceived to be improvement in the homogeneity of the finished product, almost any convenient hot deformation processing technique may be employed for consolidation purposes. Thus, hot isostatic pressing of powders may be used for making near-net shape components or parts, whereas hot extrusion may be utilized for making preforms or billets (Figure 5). On the other hand, when there is a need to preserve the initial metastable state of the powder, other methods of compaction must be employed. In dynamic compaction (Figure 5), consolidation is achieved by propagating a high intensity shock wave through the powder aggregate. Full densification is achieved when frictional heat generated between the particles is sufficient to cause surface localized melting and welding

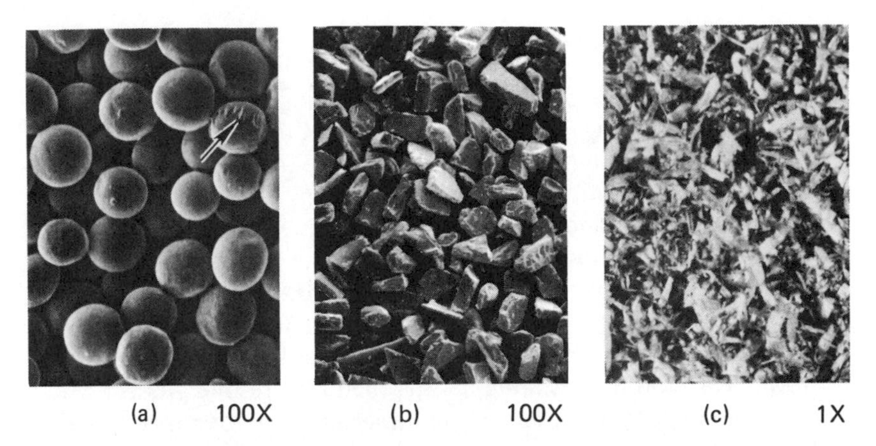

FIGURE 4 Rapidly solidified fine particulates; (a) spherical powder (product of centrifugal atomization); (b) gritlike powder (product of pulverization of melt-spun ribbon); (c) flakelike powder (product of twin roller quenching).

together of the particles. Since the heating and cooling rates are very fast, there is virtually no change in the microstructure of the material during compaction. As evidence for this it may be noted that dynamic compaction has been used successfully to produce amorphous solids from amorphous powders, splats, or ribbons.[9,10] Laser surface melting, in conjunction with continuous powder feed, has also been employed to fabricate bulk metastable structures[11] (Figure 5).

Spray Consolidation

The average cooling rate in inert gas, or centrifugal atomization, may be increased by simply allowing the atomized spray to quench out on a water-cooled chill. The resulting splats experience cooling rates of $\sim 10^6$ K/s and may be removed from the chill by scraping them off as fast as they are formed. On the other hand, thick rapidly solidified deposits may be built up by continuous superposition and bonding together of splatted particles, or, in other words, by combining particle generation, quenching, and consolidation in a single spray consolidation operation. This requires very careful control of processing variables, including melt preheat, spray deposition rate, and heat transfer characteristics. In spray rolling[12] the spray deposition rate is adjusted so that the molten droplets experience efficient splat quenching prior to completion of densification in the pinch of the rolls. In spray forging[13] the atomized spray is collected in a mold at a location in the atomizing chamber where many of the particles are in the partially solidified, or mushy, condition, which yields a preform with better than 95 percent of theoretical density. When the mold is full the hot preform is converted into a fully dense homogeneous product by closed die forging. In spray casting[14] massive ingots are formed by slowly filling the mold with a spray of fine particles generated by inert gas atomization.

Plasma spray deposition, or "plasma spraying," also combines particle melting, quenching, and consolidation in a single operation.[15,16] The process involves injection of powder particles into a high intensity plasma jet, which rapidly melts the particles and propels them toward the work-piece surface (Figure 6). Rapid quenching of the molten particles occurs when the droplets impact on the substrate. Cooling rates are typically 10^5 to 10^6 K/s, and the resulting microstructures are fine grained (~0.5 μm) and homogeneous. Conventional plasma spray deposition is normally carried out at atmospheric pressure. Typically the deposits contain oxidation products, together with some porosity due to incomplete melting, wetting, or fusing together of deposited particles.

The problem of oxidation can be minimized by shielding the plasma arc with an inert gas atmosphere. An alternative approach is to enclose the entire plasma spraying unit in an evacuated chamber, which is maintained at about 30 to 60 torr inert gas pressure by high speed pumping.[17,18] Under such "vacuum plasma spraying" conditions, plasma gas velocities are much higher (typically in the Mach 2 to 3 range), due to the higher permissible pressure ratios. Other advantages include (1) higher particle velocities, which give rise to denser deposits (often >98 percent of theoretical density), (2) broader spray patterns, which produce larger areas of relatively uniform deposition, and (3) transferred

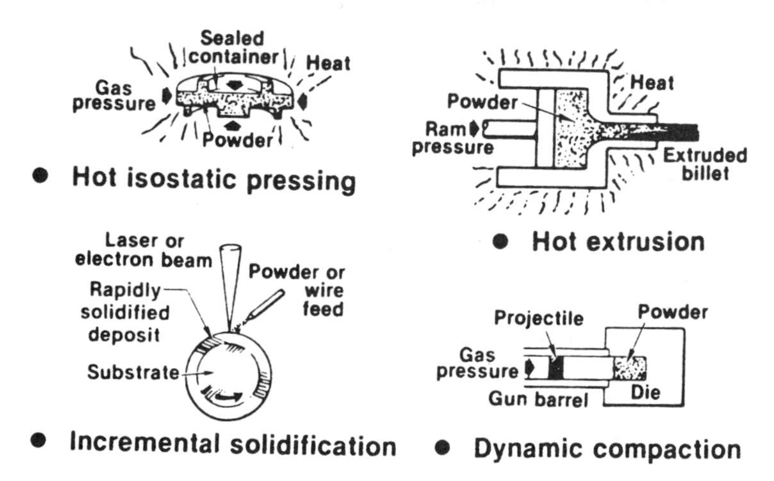

FIGURE 5 Representative consolidation methods.[3] Reprinted with permission.

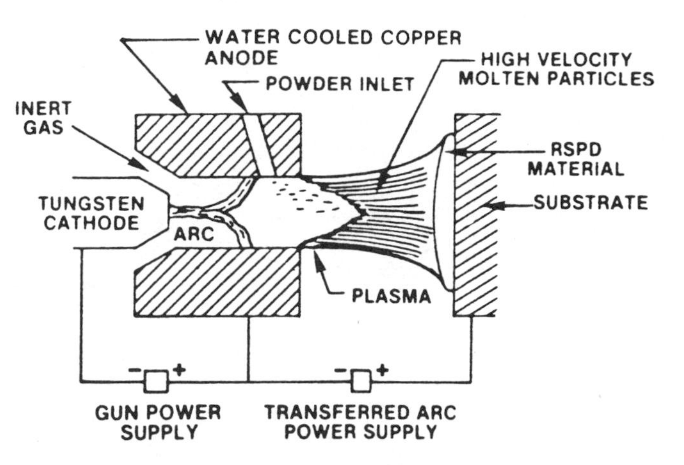

FIGURE 6 Schematic of plasma spray system.[18] In low pressure plasma spraying, the entire system is enclosed in an evacuated chamber. Reprinted with permission.

arc heating of the substrate, which improves deposition characteristics (Figure 6). In addition, the process can be automatically regulated to make controlled deposits of complex geometries at high deposition rates (up to 50 kilograms per hour [kg/hr], and in large section thicknesses (>5 centimeters [cm]), without sacrificing quality. In other words, the addition of a vacuum environment to plasma spraying has created new opportunities for near-net shape processing of bulk rapidly solidified materials.

RAPIDLY SOLIDIFIED FILAMENTS AND RIBBONS

Melt Spinning

In melt spinning[19] (Figure 3), thin filaments or ribbons are produced by forcing the melt through a small orifice directly onto the surface of a rapidly rotating copper disc, which may be water cooled for continuous operation. Current practice favors a downward directed jet (0.3 to 1.5 millimeters [mm] dia.), inclined at 15° to the disc radius, with the nozzle tip located about 3 mm from the disc surface and set back about 25 mm from its crest. The disc is typically 15 to 45 cm dia. and rotates at up to 20,000 revolutions per minute (rpm). Provided that the melt properly wets the surface of the disc, this simple jetting technique readily produces rapidly solidified filaments up to about 3 mm in width, with thicknesses that range from 25 to 100 μm. Cooling rates are in the range of 10^5 to 10^6 K/s, depending primarily on ribbon thickness. Ribbons up to

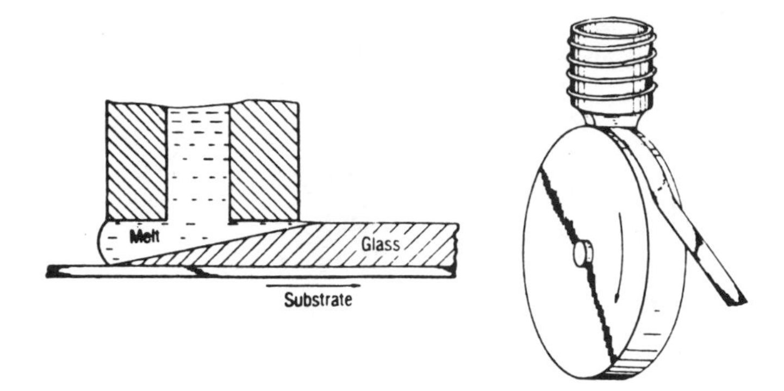

FIGURE 7 Planar flow casting process.[20]

15 cm in width can also be produced by melt spinning, but this requires careful positioning and design of the nozzle. An optimal arrangement appears to be one in which the slotted nozzle has an angled tip, which is positioned almost in contact with the surface of the rotating disc[20] (Figure 7). Such an arrangement stabilizes the melt pool that is formed under steady state conditions in contact with the disc.

The time of contact of the solidifying material on the copper chill is of decisive importance in the fabrication of amorphous ribbons. If the ribbon detaches from the disc too soon, crystallization and phase decomposition will occur during cooling in the solid state. In extended chill melt spinning[21] this is avoided by deliberately increasing the ribbon contact time by employing a spring-loaded auxiliary disc in contact with the main melt spinning disc. In centrifugal melt spinning[22] the extended chill effect occurs quite naturally, since the melt is jetted onto the inner surface of a rapidly rotating copper drum. The problem here is not to extend the contact time but rather to induce filament detachment after completion of solidification and solid state cooling. This can be done most effectively by using an inclined chill surface on the inside edge of the rotating drum, since centrifugal forces acting on the sloping surface encourage detachment of the filament. Filament contact time decreases as the slope of the inclined chill surface increases. On the other hand, the steeper the slope, the greater the tendency to form a ribbon of unequal thickness across its width.

Melt Extraction

Melt extraction is slightly different from melt spinning in that the melt source is stationary, and the edge of a rotating disc picks up the melt to form a rapidly quenched filament.[23,24] Cooling rates are somewhat slower than those attainable by melt spinning, typically about 5×10^4 K/s. The melt may be contained in a crucible, or a special arrangement may be employed that does not require a crucible, e.g., as in pendant-drop melt extraction. A typical disc for thin filament production is about 20 cm dia. and has a wedge-shaped edge. Notched or serrated discs have been used to make short fibers, or particulates. To achieve steady state processing, the melt is fed to the edge of the disc in a continuous manner by raising the molten bath in the crucible process, or by lowering the feedstock in the pendant-drop process. Electron beam melting of the feedstock is a unique feature of the pendant-drop process, which makes it particularly useful for processing reactive and/or high melting point materials.

Twin Roller Quenching

The mechanics of twin roller quenching are similar to those of melt spinning, except that a pair of counter-rotating rolls replaces a single rotating chill for the purpose of melt quenching.[25,26] Typically the melt stream is directed vertically downward between a pair of watercooled rolls, and thin filaments are formed by rapid quenching in the pinch of the rolls. In order to produce filaments of uniform thickness, the roll surfaces and shafts must be machined to close tolerances, and precision bearings must be used. Ribbons from 50 to 200 μm in thickness are formed when the rolls initially are in contact under some pressure. Thicker ribbons can be made by expanding the roll gap. Owing to the limited contact time of the solidified material with the rolls, twin roller quenching is not as efficient as melt spinning in producing amorphous materials. However, it is quite suitable for making extended solid solutions or metastable phases and has the advantage that the material can be obtained in thicker sections.

RAPIDLY SOLIDIFIED SURFACE LAYERS

Surface Melting (Glazing)

Surface modification by rapid solidification is most readily accomplished by laser or electron beam surface melting (glazing) techniques, which exploit the principle of self-substrate quenching[27,29] (Figure 3). Typically a high power density beam is rapidly traversed over the material surface so as to induce surface localized melting with high melting efficiency, i.e., melting occurs at such a high rate that there is little time for thermal energy to penetrate into the solid substrate. Under these conditions very steep thermal gradients are developed in the melt zone, which promote rapid solidification. The actual quench rate is ultimately dependent on the melt layer thickness, with cooling rates of 10^4 to 10^8 K/s readily attainable in appropriately thin sections.

Using available continuous wave carbon dioxide (CO_2) gas lasers, experience has shown that melt depths can be controlled down to ~25 μm, corresponding to an average maximum cooling rate of ~10^8 K/s. In practice, in order to exploit the microstructural/property advantages of such high cooling rates, processing must be carried out in two steps. First, the surface of the material is thoroughly homogenized by a "deep penetration" homogenizing pass, with cooling rates of ~10^4 K/s. Second, the same region of the surface is subjected to another surface melting pass, using much higher incident power density and shorter interaction time to achieve the desired higher cooling rate in a very thin surface layer.

Even higher cooling rates are possible using pulsed laser or electron beam sources, because of the higher available power densities.[30] Reproducible and controllable surface melting and quenching using pulsed sources have been achieved in layers as thin as ~1,000 angstroms (Å). Typical operating conditions in this regime of processing are power densities of ~5×10^7 watts per square centimeter (W/cm^2) and interaction times of ~10^{-8} s.

Surface Alloying

Surface alloying using high power density lasers and electron beams has also been investigated. Two distinct approaches have been evaluated: (1) preplacement of alloying material on the workpiece surface prior to melting and (2) continuous delivery of alloying material (wire, ribbon, or powder) to the interaction, or melt zone. In incremental solidification processing[11] (Figure 5), prealloyed powder is fed continuously to the interaction zone as the mandrel rotates. Thus, a much thicker, even bulk, rapidly solidified structure can be built up gradually as one deposited layer fuses to another in a continuous manner. Good interlayer bonding and epitaxial growth from layer to layer can be achieved under proper operating conditions. A critical parameter is the location of the powder impingement point with respect to the laser melt zone. Since the mandrel is rotating, feedstock impingement must occur slightly ahead of the laser beam for stable, steady state deposition. The process has great potential as a hardfacing treatment.

APPLICATIONS

Rapid solidification processing has been successfully applied to many different alloys, including nickel (Ni)-, iron (Fe)-, aluminum (Al)-, and titanium (Ti)-base alloys for structural applications and certain special alloys for magnetic and electronic applications. In order to give some feeling for the scope of these activities, a few examples will be given of developing technologies that utilize rapidly solidified fine powders, thin filaments/ribbons, or thin surface layers.

Powder Technology

Most of the effort on rapid solidification powder technology has been concerned with the further development of high performance components for gas-turbine engines, such as blades, vanes, discs, rotors, combustors, and bearings. This work has been conducted primarily at Pratt & Whitney Aircraft (P&WA) and General Electric (GE).

The work carried out at P&WA is illustrative of the more conventional approach to powder metallurgy processing, wherein powder production and consolidation are two distinct operations. Thus, at P&WA powder is produced by the proprietary "RSR" centrifugal atomization process (Figure 3), and subsequent consolidation is achieved by hot extrusion (Figure 5). After extrusion the billet is subjected to gradient annealing in order to develop a textured, columnar grained structure. Such a textured, homogeneous structure exhibits outstanding high temperature creep properties. For example, the much longer creep rupture life exhibited by RSR185 (new powder alloy), compared with directionally solidified PWA1422, is equivalent to ~70°C improvement in the metal temperature capability of this alloy. Similar improvements in creep properties, coupled with enhanced oxidation/hot corrosion resistance, have been found in derivative alloys produced by the same process.

In order to exploit such property benefits, this same processing method has been applied to the fabrication of a demonstration "radial wafer" turbine blade.[31] The wafer blade represents an advanced concept in air cooling design, which combines internal convective cooling with external transpiration, or film cooling. The sequence of steps involved in the manufacture of such a blade is shown schematically in Figure 8.

As illustrated in this figure, rapidly solidified superalloy powder is produced by the RSR process and consolidated by hot extrusion, with extrusion parameters adjusted to give an intrinsically superplastic product. The billet is rolled superplastically into sheet stock by controlling temperature and deformation rate. The resulting sheet product is pho-

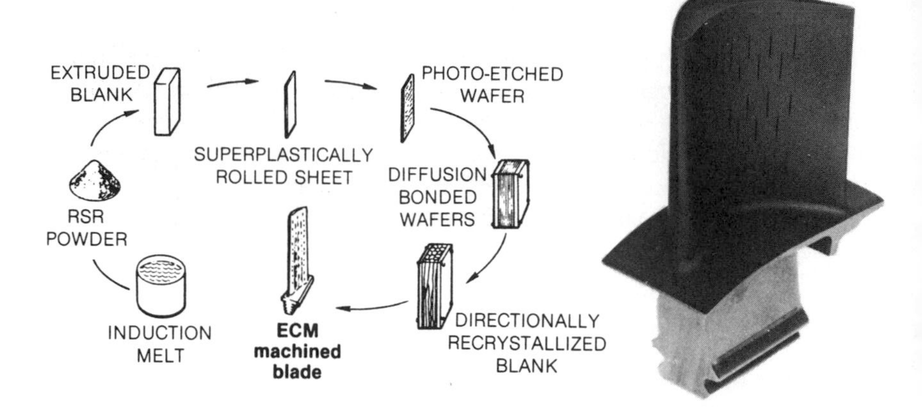

FIGURE 8 Sequence of steps involved in the fabrication of a transpiration-cooled wafer blade.[31] Reprinted with permission.

toetched to form a multiplicity of thin shaped wafers, which are diffusion bonded together in a predetermined arrangement. This is the critical step in the process, since it generates the desired network of internal cooling passages. Following directional recrystallization of the bonded structure, the actual blade profile is formed by electrochemical machining. This design concept, coupled with the higher metal temperature capability of RSR185 or a similar alloy, has the potential for increasing the turbine inlet temperature by ~350°C in the next generation of advanced engines. The experimental air cooled blade shown in Figure 8 has already been successfully tested in an advanced engine.

Using this same technology, P&WA has demonstrated significant improvements in the properties of Al-base alloys and bearing steels. Thus, certain Al-Fe-Mo alloys exhibit higher strengths than those of conventional aluminum alloys at temperatures in the range of 180°C to 350°C. Such alloys are promising candidates for integral vane and case assemblies in the cooler compressor section of the engine as replacements for the more expensive titanium alloys. Improvements in the rolling contact fatigue resistance of M50 bearing steel by rapid solidification processing also presents an opportunity for advancing the performance of high speed bearings. This beneficial effect has been related to refinement of the carbide phases in M50 steel.

In contrast to the work at P&WA, GE's effort has been concerned with combining particle melting, consolidation, and shaping in a single operation, utilizing advances in vacuum (low pressure) plasma spraying.[18] As mentioned earlier, this new technology offers a number of

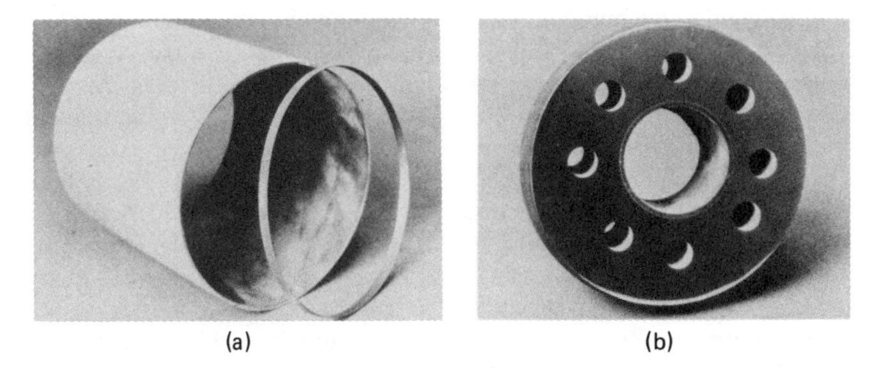

FIGURE 9 Prototype aircraft engine components made by low pressure plasma spraying of Rene 80[18]; (a) thin-walled engine combustor (0.05 cm thick × 10 cm dia.); (b) massive turbine disc (10 cm dia.). Reprinted with permission.

advantages over conventional plasma spraying, including more uniform spray patterns, higher deposit densities, and higher deposition rates. Vacuum plasma spraying of high performance coatings has become routine practice, with applications in industrial gas turbines and jet engines. Progress has also been made in the fabrication of a thin-walled combustor and a massive turbine disc (Figure 9), making use of the unique thick section capabilities of the process. Laboratory tests have shown that deposited materials, such as Rene 80, have superior resistance to thermal fatigue, which is a prerequisite for combustor applications. Currently, efforts are being made to apply this technology to the near-net shape fabrication of general engineering components, such as extrusion dies, valve bodies, pipes, casings, and sleeves.

Thin Filament/Ribbon Technology

Thin filament/ribbon technology has been developed mainly by Allied Corporation and Battelle Columbus Laboratory. Allied has favored the melt spinning process, whereas Battelle has favored the melt extraction process. Many areas of application have been identified; some have already been commercialized. Thus, today, rapidly solidified thin filaments/ribbons are being used (1) as reinforcing elements in ceramic matrix composites, (2) as interlayers for conventional and diffusion brazing, and (3) in a variety of magnetic applications.

Castable refractories are widely used in furnaces and reactors.[32] The incorporation of steel fibers (typically 0.2 to 0.4 mm^2 cross section × 20 to 40 mm long) in castable ceramics increases their resistance to thermal and mechanical cycling, thereby increasing service life. Conventional processing of steel fibers involves repeated shear-cutting of continuously drawn wires, and final embossing of the fibers to improve adhesion. The cost of processing is high, so that the 2 volume percent of fibers normally introduced into the ceramic can cost several times that of the ceramic. Battelle was first to recognize the potential for melt extracted steel fibers in this application. High aspect ratio fibers are readily and inexpensively produced by melt extraction, using a notched wheel. The resulting fibers tend to have expanded ends (dog-bone shaped), which facilitates reinforcement of the ceramic matrix. Another advantage is that melt composition is no longer limited by mechanical working considerations, so that even low grade scrap can be used for melting. Resulting savings in production costs have been substantial, and many thousands of tons of melt extracted steel fibers are used today in castable ceramics. This same process is being considered for making steel fibers for reinforcing concrete.

Diffusion brazing is a method of joining materials that combines the essential features of both conventional brazing and diffusion bonding.[33] Typically the process employs an interlayer that closely matches the composition of the workpiece, except for the addition of an appropriate melt depressant to form a low melting point eutectic. The filler material is placed between the mating surfaces of the workpiece and is permitted to alloy with it at a temperature where only the eutectic melts. Under isothermal conditions the melting point of the filler material gradually rises as the melt depressant diffuses away into the workpiece. Bonding is judged to be complete when no melt remains. Subsequent heat treatment is employed to erase all traces of the original junction. Success in diffusion brazing depends not only on good design of filler material, but also on the ability to produce the material in a usable form. A particular problem has been encountered in the preparation of thin ribbon material (25 to 50 μm thick × 2 to 5 cm wide), which is very difficult, if not impossible, to produce by conventional hot working methods because of the limited ductility of the eutectic alloy. A solution to this problem has been to prepare the thin ribbon material by melt spinning.[34] The resulting amorphous or partially amorphous material makes an attractive interlayer for diffusion brazing because it possesses moderate ductility and can easily be bent or cut to comply with complex joint geometries. Considerable success has been achieved in utilizing melt spun nickel-base alloys (boron added as melt depressant) for diffusion brazing of gas-turbine engine components, such as blades, vanes, and even entire stator rings (Figure 10).

Standard brazing alloys (e.g., those based on Ni) also contain sub-

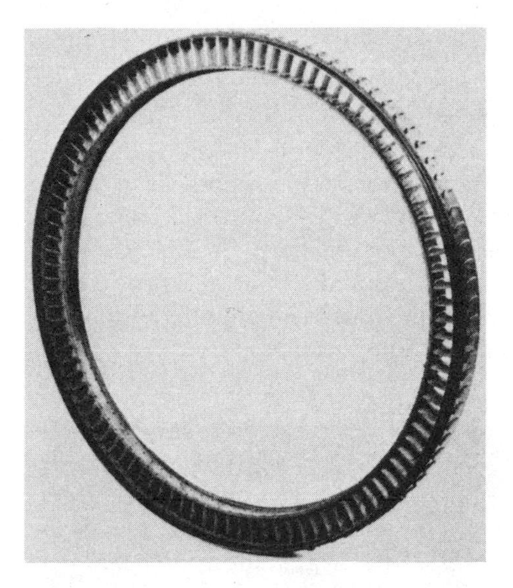

FIGURE 10 Compressor vane and case assembly for the P&WA JT8D engine (diffusion brazed with amorphous tape).

stantial amounts (~20 weight percent) of melt depressants, such as phosphorus (P), boron (B), and silicon (Si). Again such eutectic alloys are essentially unworkable but are amenable to glass formation by rapid quenching from the melt.[35] Thus, entirely new classes of brazing alloys are now available in convenient tape form. Amorphous brazing tapes have the advantages of convenience in form, chemical uniformity, and cleanliness (no binders to pyrolyze, as in conventional brazing materials), and they are relatively inexpensive to produce.

Commonly used soft magnetic alloys include Fe-3.2 percent Si for cores of power transformers and motors, and special nickel-iron alloys for electronic devices. Sheet material ~0.3 mm in thickness is used in transformer cores and motors, whereas tape 25 to 100 μm in thickness is employed in electronic devices. These materials are normally produced by a complicated sequence of rolling operations, with critical intermediate annealing steps to develop the optimal crystallographic texture and magnetic properties. Subsequent processing may involve stress relieving and coating with polymers. This complicated fabrication procedure contrasts with the simplicity of melt spinning, which produces ferromagnetic ribbon or tape directly from the melt at very high rates and at relatively low cost.

In power transformers the properties of interest exhibited by amorphous magnetic alloys, such as $Fe_{80}B_{20}$, are high saturation magnetization coupled with extremely low losses.[36] Typically, losses are down by a factor of 4 compared with the best textured iron-silicon alloy. In a finished transformer this translates into substantial energy savings over the lifetime of the installation. It has been estimated that about $200 million dollars now wasted annually as heat in transformers can be saved by substituting amorphous FeB for the best textured FeSi. Thus, there is a real incentive for pushing forward with the development of amorphous cored transformers despite certain technical drawbacks related to the thin gauge of the sheet. Prototype systems have already been fabricated (Figure 11) and are now being evaluated in actual field tests.

Various electronic device applications have been considered for metallic glasses. The first of these applications was the use of high perme-

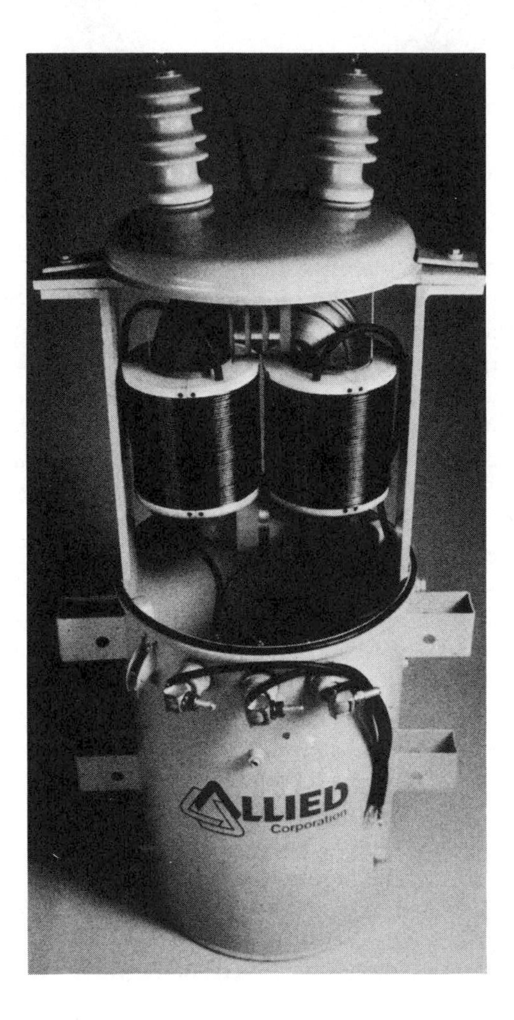

FIGURE 11 Demonstration amorphous cored power transformer.

ability amorphous alloys, e.g., $Fe_{40}Ni_{40}P_{14}B_6$, for magnetic shielding purposes. Large sheets for shielding were made by simple weaving and coating with polymers. Cylindrical shields made from these woven fabrics compared favorably in performance with conventional 80Ni-20Fe permalloy foil, except at very low fields where metallic glass loses its high permeability. The main advantage claimed for the metallic glass fabric was its ability to be formed into the required shape without altering shielding performance. Another application that takes advantage of high permeability, coupled with high electrical resistance, mechanical hardness, and resistance to corrosion and wear is for audio and video recorder heads. The preference in this application is for zero magnetostriction high-cobalt compositions with B and Si as glass formers, and twin roller quenching to produce smooth surfaces on *both* sides of the tape. Overall performance is claimed to be superior to conventional ferrites and similar materials. Other applications being considered include "stress transducers," which exploit the high stress sensitivity of the magnetic properties in amorphous alloys, and "acoustic delay lines," which make use of the very large values of magnetomechanical coupling and change in Young's modulus with applied field that are found in metallic glasses.[37] Delay lines are essential elements in all signal processing equipment.

Surface Modification Technology

Laser or electron beam surface melting (glazing) has been employed to modify the surface structure and properties of very thin edges of samples using a single pass of a sharply focused beam. On the other hand, to obtain continuous surface coverage of glazed material it has been necessary to generate a multiplicity of overlapping passes by scanning the focused beam over the workpiece surface or by indexing the workpiece with respect to a fixed beam. A laser beam may be scanned by making use of special coupled arrangements of mirrors, whereas an electron beam may be scanned by electromagnetic means. For laser glazing, a numerically controlled work station, with at least two axes of motion, is generally preferred, whereas for electron beam glazing, programmed electromagnetic beam deflection has proved to be more versatile (Figure 12).

Both laser and electron beam glazing treatments have been used to achieve beneficial modifications in the surface properties of materials. In sensitized 304 stainless steel,[38] laser glazing has the effect of resolutionizing harmful carbide phases at the grain boundaries and restores the resistance to stress corrosion cracking. In 614 aluminum bronze,[39] laser glazing homogenizes the surface, which increases its resistance to

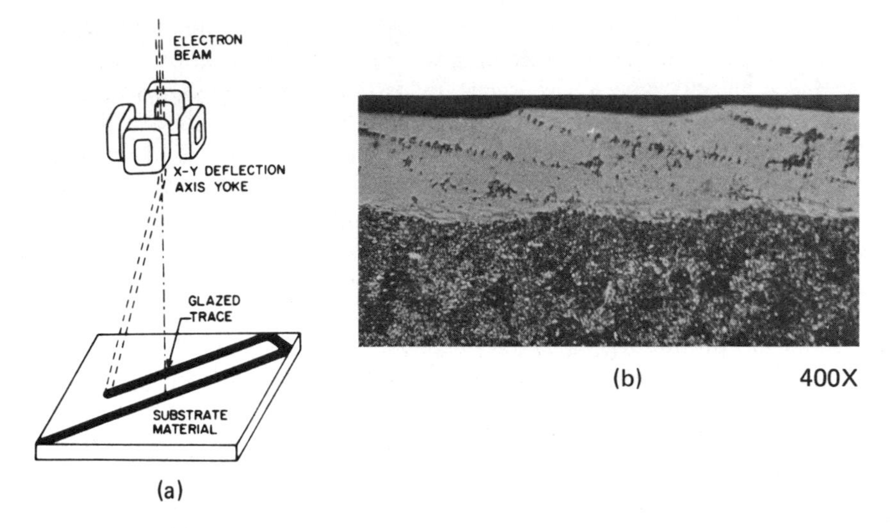

FIGURE 12 (a) Schematic of electron beam surface melting (glazing), using electromagnetic beam deflection; (b) cross-sectional view of glazed M2 steel showing overlapping passes.[29] Reprinted with permission.

corrosion in chloride solutions. In M2 high speed steel,[40] heat treatment of laser or electron beam glazed surfaces generates a uniformly fine distribution of hard carbide particles in an austenitic/martensitic matrix, which improves its cutting performance, e.g., in applications such as saw blades, drill bits, and end mills. In a pseudobinary Fe-TiC alloy,[41] electron beam glazing and tempering produce a threefold increase in the wear life in tests performed on a fully hardened M42 steel counterface material. Laser glazing has also been applied to eutectic-type alloys that are ready glass formers. Thus, amorphous surface layers have been developed on crystalline substrates in Pd-4.2Cu-5.1Si and in the technically more interesting $Fe_{40}Ni_{40}P_{14}B_6$ alloy, which exhibits exceptional mechanical properties and corrosion resistance. The high hardness and corrosion resistance of metallic glasses containing P (and chromium [Cr]), together with their ability to accept and maintain a sharp cutting edge, suggests such uses as surgeon's scalpels and even long-life razor blades.

Laser glazing in conjunction with surface compositional modification is also an area of obvious high potential. Methods of processing typically involve preplacement of alloying material (powder, electrodeposit, etc.) on the workpiece surface prior to glazing, or particle injection during glazing. Carbide particle injection into alloy substrates has been used to develop wear resistant surfaces[42] (Figure 13). Much thicker deposits

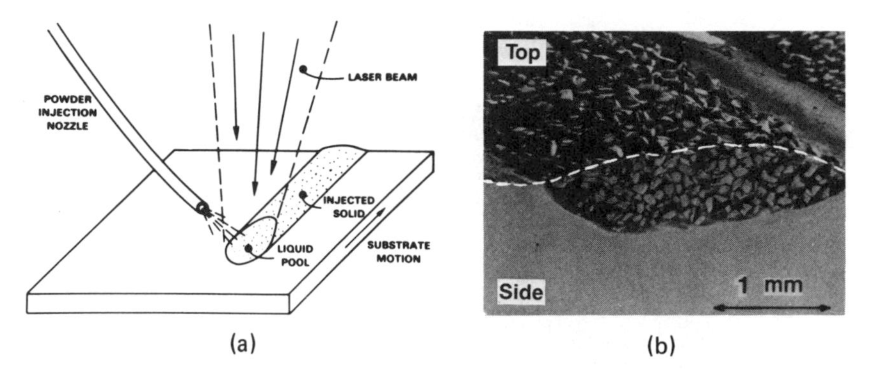

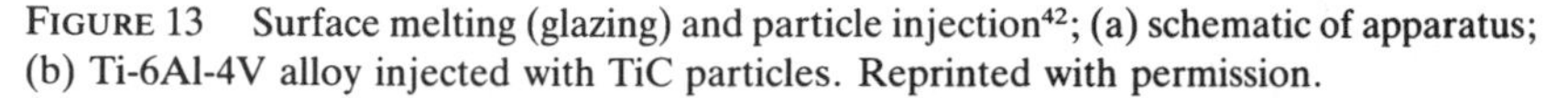

FIGURE 13 Surface melting (glazing) and particle injection[42]; (a) schematic of apparatus; (b) Ti-6Al-4V alloy injected with TiC particles. Reprinted with permission.

have also been laid down by the continuous delivery of prealloyed powder to the interaction, or melt zone (Figure 5). Surface alloying by this means is being developed for a wide range of applications, including hardfacing of valve seats, turbide blade tips, bearing surfaces, and gas-path seals. Experimental work has also been conducted on the fabrication of bulk rapidly solidified structures by incremental solidification processing. Simple axisymmetric shapes, such as a demonstration turbide disc, have already been fabricated by this process[11] (Figure 14). Typically the deposited material exhibits a pronounced columnar grained dendritic structure, with grains extending through many successive layers of material. The inherently strong tendency for epitaxial growth between

(a) (b)

FIGURE 14 Demonstration gas-turbine disc (10 cm dia.) produced by laser glazing with continuous powder feed[11]; (a) as-glazed condition; (b) after machining. Reprinted with permission.

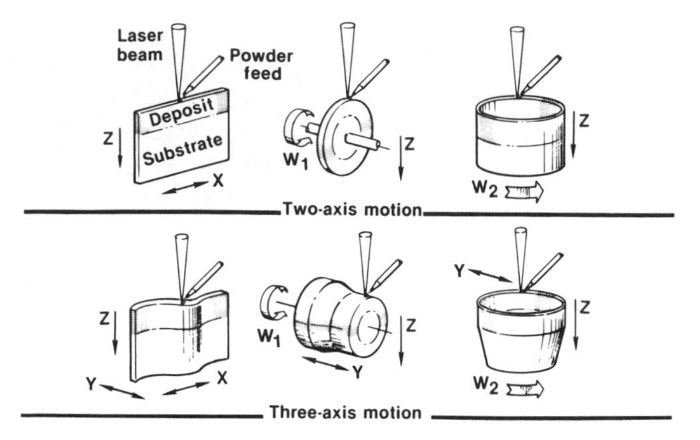

FIGURE 15 Examples of laser glazing of near-net shapes, using continuous powder feed.[11] Reprinted with permission.

layers ensures good mechanical strength at the interfaces between layers, even when the composition is deliberately changed, e.g., by changing the composition of the powder feed. Applications for this process are currently limited by the requirement that the deposited material possess good weld-cracking resistance and by the need to improve the shape-defining capabilities of the process. As indicated in Figure 15, the fabrication of more complex shapes requires the use of a numerically controlled work station, which is capable of simultaneous motion about two or three axes.

SUMMARY

The technology of rapid solidification has evolved steadily since about the mid-1970s. Today's technology includes methods for the production and consolidation of rapidly solidified fine powders, fabrication and utilization of rapidly solidified thin filaments or ribbons, and rapid solidification surface modification of materials. Powder technology has been applied to the fabrication and coating of high-performance components for gas-turbine engines. This same technology is also being applied to airframe structural materials, such as high specific strength aluminum alloys. Thin filament/ribbon technology continues to evolve impressively, with several applications already to its credit, including the use of high aspect ratio filaments as reinforcing elements in castable ceramics and of wide ribbons (tapes) as interlayers for conventional, or diffusion brazing, purposes. The anticipated use of amorphous soft magnetic alloy ribbons in the cores of power transformers and motors is also an area of high potential payoff. Surface modification technology is still in it infancy, although the benefits of rapid solidification laser or electron beam glazing treatments have been amply demonstrated. However, areas of application have been targeted for development, including hardfacing of tools, dies, and valve seats. The possible extension of this technology to bulk rapid solidification processing has also been considered.

NOTES

1. P. Duwez, R.H. Willens, and W. Klement, *J. Appl. Phys.*, *31*:1136, 1960.
2. W. Klement, R.H. Willens, and P. Duwez, *Nature*, *187*:869, 1960.
3. *Rapid Solidification Processing: Principles and Technologies II*, eds., R. Mehrabian, B.H. Kear, and M. Cohen, Claitor's Publishing Division, Baton Rouge, La., 1980 (Proc. 2nd Int. Conf. at Reston, Va., March 1980) [hereafter cited as *Rapid Solidification Processing*].
4. *Rapidly Solidified Amorphous and Crystalline Alloys*, eds., B.H. Kear and B.C. Giessen, Elsevier North Holland, New York, 1982 (Proc. Mater. Res. Soc. Meeting at Boston, Mass., Nov. 1981, Symposium F).
5. *Rapidly Quenched Metals IV*, eds., T. Masumoto and K. Suzuki, Japan Institute of Metals, 1982 (Proc. 4th Int. Conf. at Sendai, Japan, Aug. 1981).
6. H. Jones, "Rapid Solidification of Metals and Alloys," Monograph No. 8, Institution of Metallurgists, London, 1982.
7. E. Klar and W.M. Shafer, in *Powder Metallurgy for High Performance Applications*, Syracuse University Press, New York, 1972, p. 57.
8. A.R. Cox, J.B. Moore, and E.C. Van Reuth, in *Superalloys: Metallurgy and Manufacture*, eds., B.H. Kear, D.R. Muzyka, J.K. Tien, and S.T. Wlodek, Claitor's Publishing Division, Baton Rouge, La., 1976, p. 45 (Proc. of 3rd Int. Symp., Seven Springs, Sept. 1976) [hereafter cited as *Superalloys*].
9. C.F. Cline and R.W. Hopper, *Scr. Metall.*, *11*:1137, 1977.
10. D.G. Morris, *Met. Sci.*, *15*:116, 1981.
11. E.M. Breinan, D.B. Snow, C.V. Brown, and B.H. Kear, in *Rapid Solidification Processing*, 1980, p. 440.
12. A.R.E. Singer, *Met. Mater.*, *4*:246, 1970.
13. R.G. Brooks, A.G. Leatham, and G.R. Dunston, *Met. Powder Rep.*, *35*:464, 1980.
14. N.J. Grant, private communication.
15. C.W. Chang and J. Szekely, *J. Met.*, p. 57, Feb. 1982.
16. D. Apelian, M. Paliwol, R.W. Smith, and W.F. Schilling, *International Metals Review*, American Society for Metals, Metals Park, Ohio, Dec. 1983.
17. S. Shanker, D.E. Koenig, and L.E. Dardi, *J. Met.*, p. 13, Oct. 1981.
18. M.R. Jackson, J.R. Rairden, J.S. Smith, and R.W. Smith, *J. Met.*, p. 23, Nov. 1981.
19. R.B. Pond and R. Maddin, *TMS-AIME*, *245*:2475, 1969.
20. M.C. Narasimhan, U.S. Patent 4,142,571, 1979.
21. J. Bedell and J. Wellslager, U.S. Patent 3,862,658, 1975.
22. H.S. Chen and C.E. Miller, *Mater. Res. Bull.*, *11*:49, 1976.
23. R.B. Pond, R.E. Maringer, and C.E. Mobley, *New Trends in Materials Processing*, American Society for Metals, Metals Park, Ohio, 1974, p. 128.

24. E.W. Collings, R.E. Maringer, and C.E. Mobley, Tech. Rep. AFML-TR-80-70, Battelle Columbus Laboratory, Columbus, Ohio, 1978.
25. H.S. Chen and C.E. Miller, *Rev. Sci. Inst.*, *41*:1237, 1970.
26. E. Babic, E. Girt, R. Krsnik, and B. Leontic, *J. Phys. E: Sci. Instrum.*, *3*:1014, 1970.
27. E.M. Breinan, B.H. Kear, C.M. Banas, and L.E. Greenwald, in *Superalloys*, 1976, p. 435.
28. B. Lux and W. Hiller, *Prakt. Metallogr.*, *8*:218, 1977.
29. P.R. Strutt, M. Kurup, and D.A. Gilbert, in *Rapid Solidification Processing*, 1980, p. 225.
30. H.J. Leamy and G.K. Celler, in *Rapid Solidification Processing*, 1980, p. 465.
31. R.E. Anderson, A.R. Cox, T.D. Tillman, and E.C. Van Reuth, in *Rapid Solidification Processing*, 1980, p. 416.
32. J.F. Wooldridge and J.A. Easton, *Ind. Heat.*, *45*:44, 1978; *46*:42, 1979.
33. D.S. Duvall, W.A. Owczarski, and D.F. Paulonis, *Weld. J.*, *54*:203, 1974.
34. B.H. Kear and W.H. King, U.S. Patent 4,250,229, Feb. 1981.
35. N. de Cristofaro and C. Hinschel, *Weld. J.*, *57*:33, 1978; U.S. Patent 4,253,870, March 1981.
36. F.E. Luborsky, *Mater. Sci. Eng.*, *28*:139, 1977, and in *Amorphous Magnetism*, Vol. 2, eds., R.A. Levy and R. Hasegawa, Plenum Press, New York.
37. N. Tsuya and K.I. Arai, *J. Appl. Phys.*, *49*:1718, 1978.
38. T.R. Anthony and H.E. Cline, *J. Appl. Phys.*, *49*:1248, 1978.
39. C.W. Draper et al., *Corrosion*, *36*:405, 1980.
40. Y.W. Kim, P.R. Strutt, and H. Nowotny, *Metall. Trans.*, *10A*:881, 1979.
41. P.R. Strutt, B.G. Lewis, S.F. Wayne, and B.H. Kear, *Specialty Steels and Hard Materials*, eds., N.R. Comins and J.B. Clark, Pergamon Press, London, 1982, p. 389.
42. J.D. Ayers, T.R. Tucker, and R.J. Schaefer, in *Rapid Solidification Processing*, 1980, p. 212.

Exploring the Limits of Polymer Properties: Structural Components From Rigid- and Flexible-Chain Polymers

ROGER S. PORTER

It has been seven decades since Baekeland commercialized the first plastic produced from small molecules. Since then, synthetic polymers have come to fill an array of needs, many generated by the very availability of these extraordinary materials.[1] The increase in production and use of polymers has been spectacular. In the United States alone production in 1981 exceeded 24 million metric tons. In volume this exceeds the production of steel. Polymer production represents more than $100 billion of value added by manufacture and involves the employment of 3.3 million people. Polymers thus represent a large, rapidly expanding, and significant class of materials of importance to both the economy and national security.

The rapid growth in use of polymers and their substitution for other materials has led to the design and evaluation of new polymer products. Short-term polymer tests designed to accelerate degradation have been commonly used. Frequently these tests are not adequate for more than the ranking of performance under a single set of conditions. As a result, polymers may fail prematurely and unexpectedly in use. If the advantages of polymers are to be fully exploited through innovative use, meaningful service-life prediction and nondestructive characterization methods must be further developed. This need has been recognized in

The introduction to this paper is based on "Opportunities and Needs for Research on the Performance of Polymers" by R. K. Eby, Chief, Polymer Science and Standards Division, National Bureau of Standards, Washington, D.C.

at least six studies.[1-6] For example, a National Research Council report[1] lists research opportunities in polymer science, and one-half of these deal with accelerated tests for long-term behavior, viz, failure and degradation of polymers.

From a positive perspective, this paper first illustrates several recent polymer applications, particularly those in construction and transportation. Among advanced uses for bulk polymer is the Bethlehem Steel Corporation process for making biaxially oriented polypropylene that is tough enough to stop a bullet. More generally, the performance of polymers is being enhanced with increasing sophistication by reinforcement with glass, or, better, by reinforcement with polymer fibers, and finally by the ultimate generation of self-reinforcing polymer. The paper thus concludes with a brief description of major and recent advances that can provide the next generation of high-performance polymers. The major opportunities are still in the future for polymer applications and for substitution of other materials. This is well recognized abroad, as is seen in the section below on "The Japanese Challenge."

RECENT APPLICATIONS

The lighter weight of reinforced plastics, with consequent savings in fuel, is a major factor in the switchover from metals to reinforced plastics in automobiles and aircraft. New fabricating techniques that make possible the production of composite parts at a lower cost than for all-metal parts are also playing a growing role in this change. Among the new and demanding applications is a pultruded graphite-reinforced helicopter windshield post that is part of a Department of Defense contract for a prototype helicopter. The publicized Lear Fan Turboprop makes such use of carbon/epoxy composites that only items such as the engine, landing gear, and wing-attach fittings are metal. Wing-tip fuel tanks for the F-18 fighter plane are a complex composite that withstands fire, impact, and even bullets that destroy all-aluminum tanks. (With reference to recent applications, *see* Society of the Plastics Industry, *News of 1983*.)

In automobiles polymers have been used for many years for decorative, nonstructural purposes. Present considerations are to use plastics in more stress-critical components, such as hood, trunk lid, and structural frame, for weight saving. Plastics will, in such evolutionary fashion, find their way more and more into critical automobile applications requiring strength and stiffness. The pace at which this occurs will depend on factors such as the development of a data base for engineering design and on the ability of engineers to use the data. While it may seem that we are in the "Plastic Age," we are just beginning to see the options. In 1977 Ford planned for a prototype car with body, chassis, and power-train components made of graphite-fiber composites. This project was undertaken to demonstrate the potential of graphite-fiber-composite technology for construction of a light-weight car with good fuel economy, yet retaining the performance, interior space, and comfort of larger vehicles. The completed experimental vehicle weighed 2,504 pounds, some 1,246 pounds less than a 1979 production Ford LTD equipped with a 351, 5.0-liter CID engine. Only the power train, trim, and some chassis components were not converted. Even most of these (e.g., engine, brakes, and transmission) could be downsized for secondary weight reductions.

THE JAPANESE CHALLENGE

The world position of nations is influenced by technology. In contemporary competition, military materiel has become the science and engineering of materials. This has been dramatically illustrated by the delivery of steel from Japan to Pittsburgh at a favorable price and quality. Our polymer developments are also being challenged in Japan. The following stark example is from a translation of *Nikkei Sangyo Shimbrun* (July 27, 1982):

> The Ministry of International Trade and Industry (MITI), the synthetic fiber industry, and academic institutions are to engage in the joint development and practical application of "the third generation fiber" which will have more than twice the high tenacity, high modulus, low elongation of the present fibers. With government subsidy, MITI has designated, effective fiscal year 1983, this next generation research/development program. The industrial infrastructure plans to allocate 3 billion–5 billion yen in funds, with practical application targeted five years ahead. The new fiber is expected to replace nylon and carbon fiber and expand the area of fiber demand. MITI and Japanese industry look forward to this development project as a conclusive factor for the revitalization of the fiber industry, now suffering from recession, and for increasing the value-added for polymer products.

At present the closest fiber to this third-generation fiber is Du Pont's Aramid (a grade known as "Kevlar-49"), with a high tenacity of ~28 grams per denier, ~3.6 gigapascals (GPa), the world's most tenacious commercial fiber, but with a maximum modulus inferior to that of carbon fiber and ultradrawn polyethylene. The Japanese project may be the world's first development project in the area of new materials that is the equal in importance with advances in electronics and biotechnology. In typically Japanese fashion, this research is to be conducted jointly within overlapped government, industry, and academic circles, and under government subsidy.

RECENT RESEARCH TOWARD HIGH-MODULUS POLYMERS

Polymer researchers have approached the problem of making the strongest possible polymers in two diverse ways: (1) by chemically constructing polymers with rigid and linear backbone chains and (2) by processing conventional flexible-chain polymers in ways that result in a transformation of the internal structure and properties. Chemical construction of rigid macromolecules has been approached by syntheses leading to parasubstituted aromatic rings in the polymer backbone. In general, these polymers cannot be processed by means of conventional polymer techniques; however, some industrial examples, viz, Du Pont Kevlar and Monsanto X-500, have been solution-processed into fibers of very high strength.

In the second category, flexible-chain polymers are converted into highly oriented and chain-extended conformation, with substantially increased tensile moduli, by drawing from dilute flowing solution or from a gel state or by extruding a supercooled melt by solid-state extruding or by drawing below the polymer melting point under controlled conditions.

FLEXIBLE-CHAIN POLYMERS

New and successful drawing techniques for flexible-chain polymers have been recently developed by workers in several countries—in the United States at the University of Massachusetts and elsewhere, and in Japan. It has been found possible, for example, to ultradraw single-crystal mats of ultrahigh-molecular-weight polyethylene (UHMWPE). By the principal deformation technique of solid-state coextrusion, draw has been achieved even at room temperature and at up to 130°C, i.e., just below the melting point. Moreover, the resulting stable extrudate exhibits extreme orientation. Multiple drawing by repeated coextrusion at 110°C produces an extrudate of UHMWPE with a draw ratio (DR) of 110 and a tensile modulus of 100 GPa. An even higher DR has been achieved by a combination of solid-state coextrusion followed by tensile drawing at controlled rate and temperature. The maximum achieved for the present by this drawing combination is a DR of 250. This superdrawn sample has a tensile modulus of 222 GPa, which is about twice the highest previously reported room-temperature experimental value (110 GPa) for polyethylene. Figure 1 summarizes some of these new results.

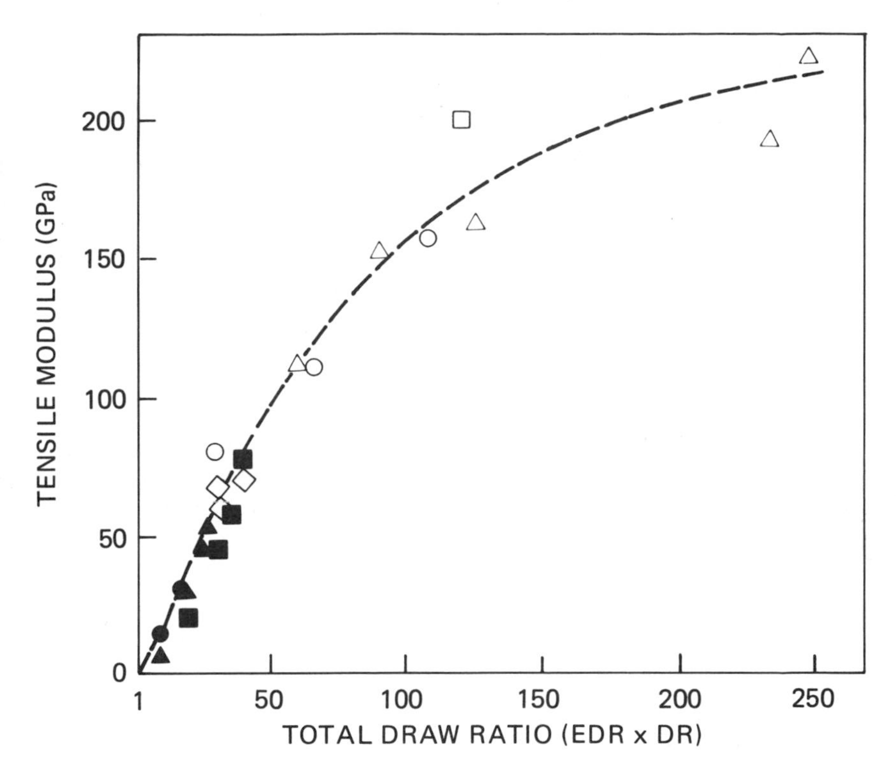

FIGURE 1 The tensile modulus of high-density polyethylene increases markedly with draw attained by linear extension. A draw ratio (DR) of 10 means extending 10 times by solid-state extrusion (EDR) and followed with tensile pulling.

RIGID-ROD POLYMERS

Carbon and graphitic fibers produced from polymeric precursors exhibit some of the highest performance characteristics of materials available to date. Indeed, such fibers have been extensively investigated over the last two decades owing to their high-temperature stability and exceptional mechanical properties. Commercially available fibers possess tensile moduli of up to 690 GPa along with tensile strengths of 2.2 GPa.[7] Such fibers, however, are quite brittle, which may limit their use in certain applications. Also, to produce carbon and graphitic fibers, extreme processing conditions are required, leading to high production and product costs. The electrical conductivity of these fibers is also not always desirable in application. Thus, there still exists a need for additional high-performance polymers. Indeed, research continues in related areas, and a sizable activity concerns extended-chain and rigid-rod polymers possessing high-performance characteristics.

Fibers produced from lyotropic liquid crystalline solutions of extended-chain polymers have not only achieved desirable high-performance characteristics but have become successful engineering materials through the development of conventional wet spinning techniques for their manufacture. Both Monsanto[8] and Du Pont[9] have had success in developing high-modulus/high-strength fibers based on wholly aromatic polymers that possess a rodlike character derived from steric effects; however, only Du Pont has pursued commercial development (Kevlar). Even here there are interesting but disconcerting limitations in compression and shear (see Figures 2 and 3). Remarkably, stretching after compression produces a virtually restored high-modulus Kevlar-49.

Initial success in producing fibers from extended-chain macromolecules has encouraged further investigation of rigid-rod polymers. A sizable research effort sponsored by the U.S. Air Force Wright-Patterson Materials Laboratory and the U.S. Air Force Office of Scientific Research (Ordered Polymer Research Program)[10] is currently evaluating the nature of novel rigid-rod macromolecules. The University of Massachusetts is playing a major part in this activity. To our knowledge, it is the only research of this type in the Western world; it is described briefly below.

The goals of the Air Force Ordered Polymer Research Program have focused on extended-chain, aromatic heterocyclic molecular structures. Three of the polymers synthesized as part of this program are a poly-

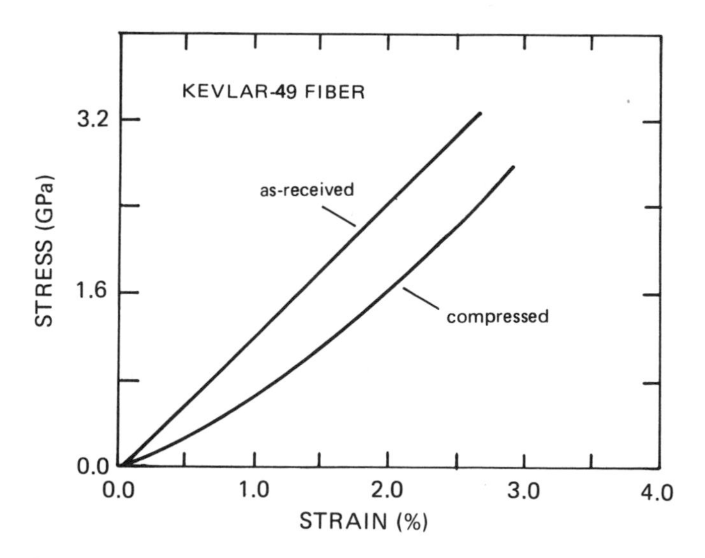

FIGURE 2 Tensile properties of as-received and compressed Kevlar-49 fiber.

FIGURE 3 Micrographs comparing Kevlar-49 fibers before (left) and after (right) axial compressive failure, showing helical kink bands.

benzimidazole, PDIAB[11]; poly-(p-phenylene benzobisoxazole), PBO[12]; and poly-(p-phenylene benzobisthiazole), PBT.[13,14] Of these structures the PBT polymer offers the best thermal and oxidative stability. For these reasons, since 1978 the emphasis has centered on its development. PBT is soluble only in strong acids.[14,15] The viscosity of such solutions passes through a maximum with increasing polymer concentration, indicating formation of a lyotropic liquid crystal phase. The ability of PBT solutions to be spun from this mesophase with formation of high-modulus/high-strength fibers has been demonstrated. Heat-treated fibers with moduli of 300 GPa and strength of 3 GPa have been routinely produced.[16] These fibers are highly anisotropic and, like Kevlar, are considerably weaker in shear and compression than their graphite or glass competitors, as illustrated above. However, this characteristic also gives these fibers the amazing flaw insensitivity and non-brittle-type behavior in compression and shear that permits their use in applications such as bulletproof vests. Thus we now have seen by at least two diverse routes—flexible and stiff chains—that crystallizable polymers have been developed into structures of both extraordinary tensile and impact properties. This is leading to a range of applications well beyond the innovations described above. The potential for polymer applications remains unbounded.

NOTES

1. *Polymer Science and Engineering: Challenges, Needs and Opportunities,* Report of the Ad Hoc Panel on Polymer Science and Engineering, National Research Council, Washington, D.C., 1981.
2. *Organic Polymer Characterization,* NMAB 332, National Materials Advisory Board, National Research Council, Washington, D.C., 1977.
3. *Polymer Materials: Basic Research Needs for Energy Applications,* CONF-780643, U.S. Department of Energy, Washington, D.C., 1978.
4. *Morphology of Polyethylene and Cross-linked Polyethylene*, Workshop proceedings; EL-2134-LD, Electric Power Research Institute, 1981.
5. *Organic Matrix Structure Composites: Quality Assurance and Reproducibility,* NMAB-365, National Materials Advisory Board, National Research Council, Washington, D.C., 1981.
6. *Materials for Lightweight Military Combat Vehicles*, NMAB-396, National Materials Advisory Board, National Research Council, Washington, D.C., 1982.
7. W. Bruce Black, "High Modulus/High Strength Organic Fibers," *Annu. Rev. Mater. Sci., 10*:311, 1980.
8. W.B. Black and J. Preston, eds., *High Modulus Wholly Aromatic Fibers*, Marcel Dekker, New York, 1973.
9. H. Blades, U.S. Patent 3,869,430, "High Modulus, High Tenacity Poly(p-Phenylene Terephthalamide) Fiber," assigned to Du Pont, 1975.
10. T.E. Helminiak, "The Air Force Ordered Polymers Research Program: An Overview," *Am. Chem. Soc. Org. Coat. Plast. Prepr., 4*:475, 1979.
11. R.F. Kovar and F.E. Arnold, "Para-Ordered Polybenzimidazole," *J. Polym. Sci., Polym. Chem. Ed., 14*:2807, 1976.
12. T.E. Helminiak, F.E. Arnold, and C.L. Benner, "Potential Approach to Non-Reinforced Composites," *Polym. Prepr., Am. Chem. Soc., Div. Polym. Chem., 16*(2):659, 1975.
13. J.F. Wolfe, B.H. Loo, and F.E. Arnold, "Thermally Stable Rod-like Polymers: Synthesis of an All-*Para* Poly(Benzobisthiazole)," *Polym. Prepr., Am. Chem. Soc., Div. Polym. Chem., 19*(2):1, 1978.
14. J.F. Wolfe, B.H. Loo, and F.E. Arnold, "Rigid Rod Polymers. 2. Synthesis and Thermal Properties of Para-Aromatic Polyamides with 2,6 Benzobisthiazole Units in the Main Chain," *Macromolecules, 14*:915, 1981.
15. E.W. Choe and S.N. Kim, "Synthesis, Spinning and Fiber Mechanical Properties of Poly(p-Phenylene Benzobisthiazole)," *Macromolecules, 14*:920, 1981.
16. S. Allen, "Mechanical and Morphological Correlations in Poly(p-Phenylene Benzobisthiazole) Fibers," Ph.D. thesis, University of Massachusetts/Amherst, 1983.

ACKNOWLEDGMENT This review was prepared initially with Richard J. Farris for presentation at the Workshop on Substituting Non-Metallic Materials for Vulnerable Minerals sponsored by the National Science Foundation, Washington, D.C., June 27-28, 1983.

High-Technology Ceramics

ALBERT R. C. WESTWOOD AND JAN P. SKALNY

The excitement in the field of ceramics these days is referred to in Japan as "ceramic fever." It relates to the intriguing prospects of having this class of solids finally live up to its potential in terms of strength at high temperatures, resistance to environmental degradation, and low cost.

Of course, these are not traditional ceramics—those beautiful, usually rather fragile examples of the potter's art made from naturally occurring substances such as clays, talc, and feldspars (Figure 1).[1] Early materials technologists were able to increase the durability of such products somewhat by applying glazes, substances formulated both to improve appearance and to have a coefficient of thermal expansion smaller than that of the substrate ceramic. On cooling, the substrate placed the glaze into compression, making it more difficult for minor scratches to develop into catastrophic cracks. Unfortunately, traditional ceramics suffer from two almost insurmountable disadvantages—first, the inevitable presence of voids and microcracks at phase boundaries that serve as crack nuclei and, second, the absence of any means of stopping a crack once it gets started.

The "high-technology ceramics" (HTCs) now beginning to emerge from R&D laboratories are designed to reduce or circumvent both of these failings. But to develop these, ceramists literally have had to start again from the beginning. Instead of using natural but impure and irreproducible starting materials, they are now using chemically pure substances, such as alumina, silica, carbon, and nitrogen. Their approach

FIGURE 1 Grecian vase, circa seventh century B.C., prior to restoration.[1] Reprinted with permission.

is similar to that used to make semiconductor materials for electronic applications, with great emphasis on control of composition and structure.

Because the strength of a ceramic solid is inversely related to the square root of the size (d) of its largest flaw, and since d usually depends directly on the size of its component "grains," the modern ceramist prefers to build up ceramic shapes from chemically homogeneous particles, typically 1 micrometer (μm) or less in diameter and produced by a variety of essentially chemical routes. Such particles are compacted and then heated to sinter them together into a ceramic solid, but at much lower temperatures than those used in the past. There are two reasons for this. First, reduced temperatures and times are sufficient because the diffusion distances needed to accomplish densification are smaller. Second, it is necessary to avoid thermally induced growth of the small particles because, as noted above, structures containing larger grains are weaker.

The approach described may be contrasted with that for traditional ceramics in which substantial masses of constituents are reacted at relatively high temperatures and the resulting clinker is ground down to inhomogeneous particles of perhaps 5 to 20 μm in diameter that are then sintered back together, again at relatively high temperatures. Grinding is avoided in the new approach not only because it is inefficient, costly, and time-consuming, but also because it introduces impurities from the grinding media. Ideally, comminution, or grinding, is now used only to break up loosely bound clusters of fine particles or to add strain energy to them to accelerate subsequent sintering reactions.

Developments leading to the generation of HTCs, or, as the Japanese prefer to call them, "fine ceramics," are occurring in three areas: the production of powders, of shapes, and of the important property of toughness. Some comments on progress in each of these areas and on possible applications for these new ceramics are presented next.

POWDER PRODUCTION

Various approaches are being pursued to produce chemically homogeneous, submicron-dimensioned particles of ceramics such as TiB_2, ZrO_2, SiC, or of tough ceramic alloys such as partially stabilized ZrO_2 (PSZ).

The routes followed involve either liquid- or vapor-phase chemistry. In the first case, this usually involves production of a colloidal suspension and subsequent removal of the solvent. Unfortunately, if simple drying by heat or evaporation is used, coarse crystals or agglomerates often result. Accordingly, other techniques, such as spray- or freeze-drying, are generally used. An alternative technique that has attracted some attention is the solgel approach. This usually involves three steps: (1) producing a concentrated dispersion of a metallic salt in a dilute acid (the sol); (2) adjusting the pH, adding a gelling agent, and evaporating the liquid to produce a gel; and (3) calcining the gel under carefully controlled atmospheric conditions to produce fine particles of the requisite ceramic. This approach has proved especially useful for oxide-based ceramics, e.g., Al_2O_3, ZrO_2, and ferrites.

Another popular method for making colloidal dispersions is via the hydrolysis of metal alkoxides, these being the products of reaction between alcohols and metal oxides. One advantage of this approach is that the alkoxides—and hence the product oxides—can be purified by distillation. Also, the precipitated hydroxides tend to be uniform, spherical, submicron particles. Excellent packing densities can be obtained by the use of conventional colloid chemistry procedures.

An example of TiO_2 ceramic solid produced by calcining ethoxide-derived precursor particles at 1050°C is shown in Figure 2(b).[2] The solid is >99 percent dense, and the grain size is ~1.2 μm. The precursor

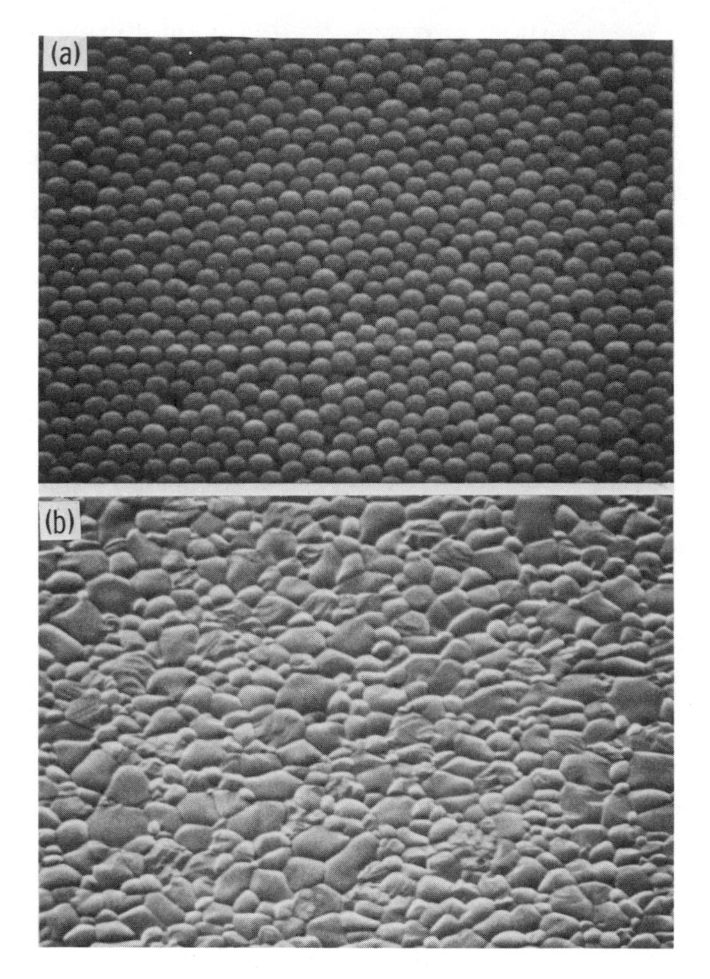

FIGURE 2 (a) TiO_2 powder, about 0.35 μm in diameter, produced by the alkoxide route. (b) TiO_2 ceramic of >99 percent theoretical density made from powder above. Average grain size is ~1.2 μm.[2] Reprinted with permission.

particles from which this solid was produced are shown in Figure 2(a). They are ~0.35 μm in diameter. The sintering temperature required in this case was some 300°C below that conventionally used for TiO_2. Lead zirconium titanate (PZT) made via a butoxide route likewise can be sintered at 950°C instead of the conventional 1300°C, and alkoxide-produced ZrO_2 can be sintered at 1200°C to 1300°C instead of the normal 1500°C to 1700°C.

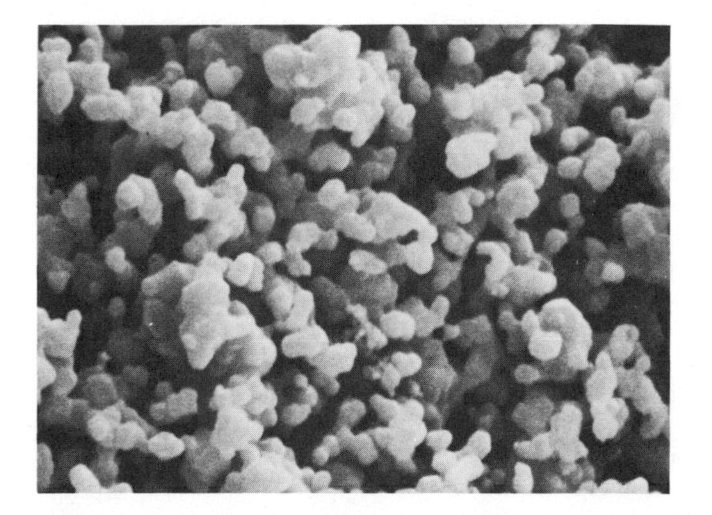

FIGURE 3 SiC particles, ~0.5 μm in diameter, produced by reacting SiO vapor and C particles at 1600°C.[6]

Alternate routes lie in reactions that occur in the vapor phase. Various energy sources, including plasmas and lasers, have been used to promote such reactions. In the latter case researchers at MIT and elsewhere are making fine particles of silicon (Si), SiC, and Si_3N_4 from gaseous reactants such as SiH_4, NH_3, and C_2H_4, using a CO_2 laser. The powders produced typically are <0.1 μm in diameter.[3] In another vapor-phase route, water vapor is reacted with aerosol droplets of alkoxides to produce either pure or mixed oxide powders of less than 1 μm diameter.[4]

While technically elegant, the cost-effectiveness of such approaches has yet to be established, and sometimes it can be useful to revisit an old approach with the perspective of improved mechanistic understanding. To illustrate: It is now recognized that the mechanism by which SiO_2 and carbon interact at high temperatures to produce SiC involves the intermediate production of gaseous SiO.[5] This substance adsorbs onto the carbon particle and reacts with it to produce SiC and CO. If the carbon particles are small, the reaction can be controlled in such manner that the size and shape of the SiC particles are essentially determined by that of their carbon precursor.[6] SiC particles ~1 μm in diameter have now been produced in this manner at temperatures as low as 1600°C, some 500°C below the traditional processing temperature (see Figure 3).[6] No subsequent grinding is necessary. This approach, when scaled up, should permit the production of submicron-sized SiC for sale at $5 to $10 per pound.

PRODUCING CERAMIC SHAPES

Two principal steps—forming and consolidation—are involved in making a ceramic shape, and improved processes are being developed for both. The trend in forming is toward utilization of an approach standard in the plastics industry but actually developed for ceramic shape production in the 1930s, namely, injection molding. In this process the ceramic powder, together with a low-melting-point binder, is heated to 150°C and then injected into a mold under a pressure of about 5,000 pounds per square inch (psi). The part so produced is then heated to ~200°C to evaporate most of the binder, the product being a weakly bonded "green" (unsintered) shape. Sufficiently complete removal of the binder to then permit the eventual production of a component of close to theoretical density can take several days, although NGK in Japan has recently reported reducing this time to about one day.

Conventionally, the green shape is then heated at some elevated temperature in a controlled atmosphere to sinter the ceramic particles together and to produce the final part. During the past few years, however, the technique of hot isostatic pressing (called HIPing) has been introduced to further densify the sintered part and so minimize the number of fracture-initiating voids remaining. This technique can produce complex ceramic shapes of <2 percent porosity and with pore sizes sufficiently small that room-temperature tensile strengths of 50 to 100 kpsi (thousand pounds per square inch) are achievable.

Unfortunately, the equipment required for "HIPing" is rather expensive—an industrial unit might cost $1 million or so. The process cycle also is slow. Moreover, ceramics such as SiC are intrinsically difficult to sinter, requiring the use of sintering aids the presence of which can be detrimental to subsequent high-temperature performance. Accordingly, alternative approaches are now being investigated, among which are high-pressure (200 atmosphere) gas sintering,[7] shock processing, and "rapid omnidirectional compaction," or "ROCing."

Shock processing, using explosives, is a collective term for three possible procedures: (1) shock loading to enhance sinterability, (2) shock compaction (preferably without binding agents) followed by conventional sintering, and (3) "one-shot" compaction and sintering.[8] Because of the substantial pressures (to 1 megabar [Mbar]), high local temperatures, and short reaction times involved, shock processing permits the production of nonequilibrium phases and the use of novel combinations of materials to provide preferred properties, e.g., toughness, low electrical conductivity, or enhanced catalytic efficiency. In just the past year, shock compaction has been used to produce high-density test pieces from Si_3N_4, Al_2O_3, ZrO_2, TiC, and TiB_2, the latter material exhibiting

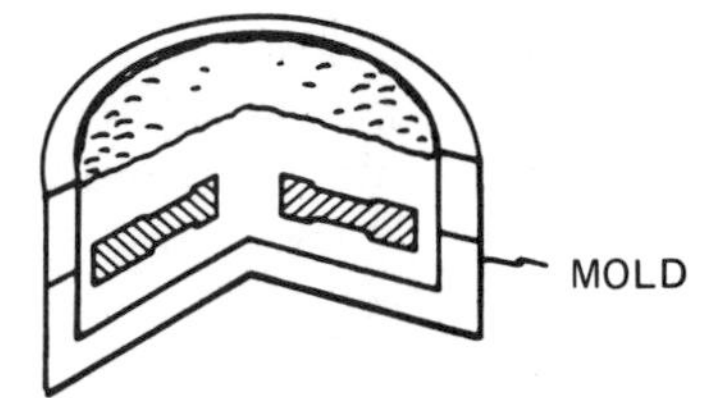

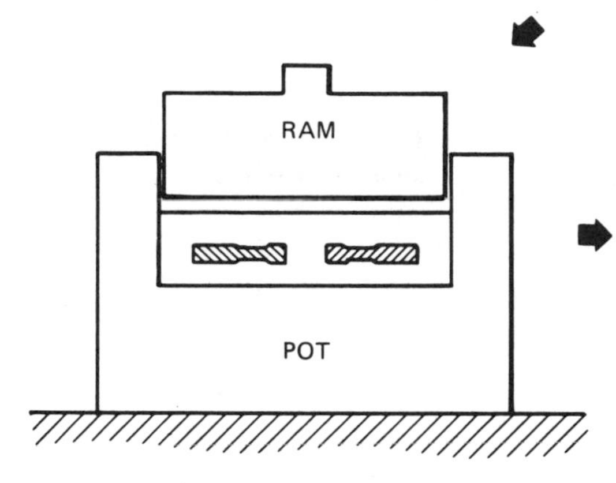

FIGURE 4 Elements of Rapid Omnidirectional Compaction (ROC) process.[10] Reprinted with permission.

the respectable fracture toughness of ~9 $MPam^{1/2}$ (megapascals square-root meter).[9]

Another process used to date only with metals but exhibiting promise for the future production of near-net shapes in ceramics makes use of a "fluid"-containing die to transfer the pressure generated by a forging press isostatically to the part. This process, termed ROCing, is illustrated in Figure 4.[4] The "fluid" used can be mild steel, a Cu-10 percent Ni alloy, or a variety of proprietary glasses, the appropriate medium being one that is very plastic or molten at the forging temperature. Pressures of order 100 to 150 kpsi can be applied to the green compact, these being several times that typical for a HIPing operation. The availability of such high pressures should permit the use of reduced forging temperatures, producing finer-grained, stronger products.

After cooling, the solidified die medium must be machined, melted, or fractured away from the shape. Even so, this technique appears to provide several advantages over HIPing, not the least of which is the use of conventional, readily available forging presses, and the brief in-press cycle time, one stroke of the press being sufficient. In comparison, a HIPing cycle usually takes several hours.

THE DEVELOPMENT OF TOUGH CERAMICS

Over the past few years, materials scientists have sought to circumvent the intrinsic fragility of ceramics by reducing the size and concentration of preexisting flaws through the use of ultrafine particles and compacting processes capable of producing components of near-theoretical density, as just discussed, and by introducing into the ceramics a variety of synthetic crack-retarding entities, such as phase-transforming particles, fibers, and distributions of cracks (see Figure 5).

The critical factors determining the relative efficiency of particulate

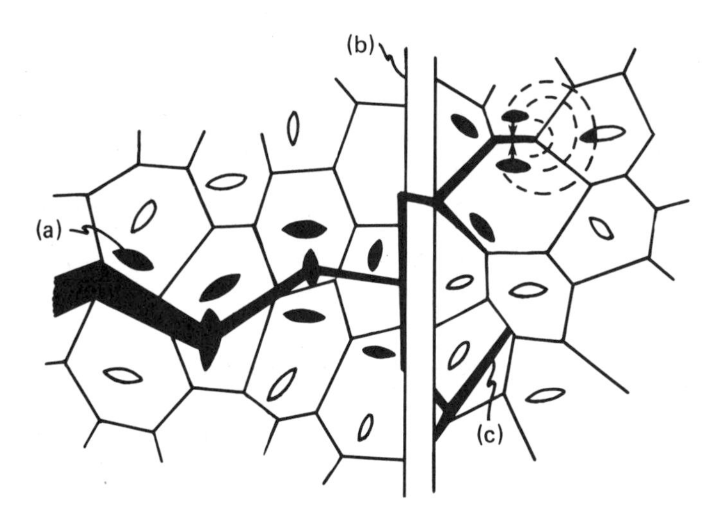

FIGURE 5 Crack-retarding entities used to produce toughness in ceramics; (a) phase-transforming particles, e.g., ZrO_2 (tetragonal) → ZrO_2 (monoclinic); (b) fibers with weak fiber-matrix interfaces; and (c) other cracks.

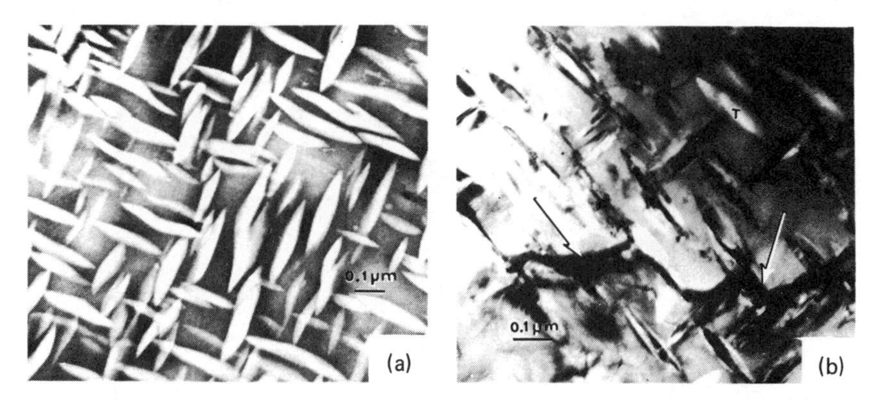

FIGURE 6 Partially stabilized ZrO_2 (a) unstressed, showing coherent tetragonal precipitate particles; (b) stressed by indenter, revealing transformed monoclinic particles near indentation (arrows) and untransformed tetragonal particles elsewhere.[12] Reprinted with permission.

crack-retarding entities are the size, concentration, and spacing between the dispersed particles or fibers, and the differences in mechanical properties between them and the matrix.[11] Essentially, the smaller and closer the particles and the bigger the differences in properties the better.

The toughening of otherwise brittle inorganic solids by the addition of fibers has long been practiced, e.g., in asbestos cement. But it is now recognized that to obtain any substantial increase in the toughness of ceramics, fiber diameter and spacing should be <10 to 50 μm. Homogeneity of distribution also is very important.

The most interesting current exploitation of these principles occurs in the developing class of "transformation-toughened" ceramics. In these solids dispersed small particles of some metastable phase are transformed crystallographically when the strain field of a crack passes through them. Some of the energy of the crack is absorbed thereby. If the particles also increase in volume, they can apply a compressive stress to the crack tip, reducing its effective driving force. Further crack-retarding interactions occur at the particle-matrix interface and within the particle itself.

The best-known example of this behavior occurs in partially stabilized ZrO_2 (PSZ). In this case, a two-phase ZrO_2 is produced by partially stabilizing the tetragonal ZrO_2 phase by additions of up to 10 percent Y_2O_3, MgO, or other oxides. A typical structure is shown in Figure 6.[12] Note the relatively high volume, crystallographic orientation, and small size (0.5 to 2.0 μm) of the tetragonal phase. When a crack cuts through this material, the tetragonal phase transforms locally into a monoclinic

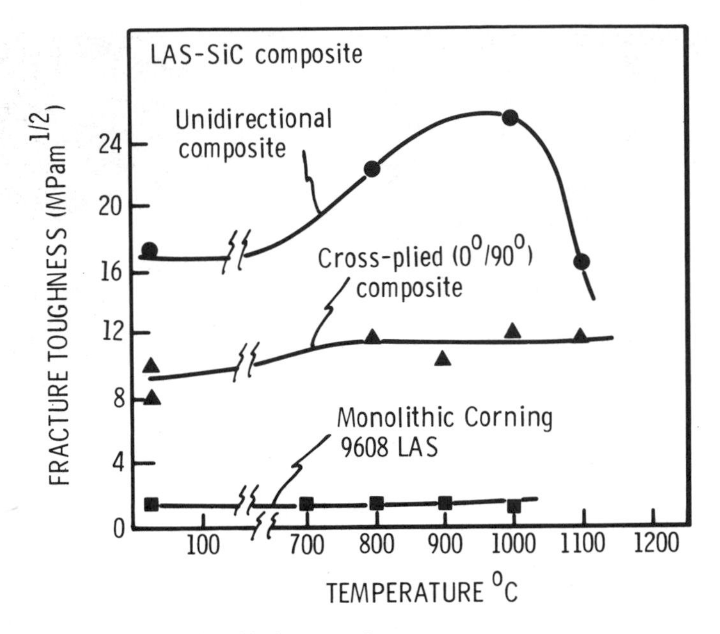

FIGURE 7 Relationship between fracture toughness and temperature for a lithium-alumino-silicate glass—50 percent SiC fiber composite.[16] Reprinted with permission.

structure, and toughening occurs by the mechanisms described above. Polycrystalline PSZ can exhibit room-temperature strengths of 50 to 150 kilograms per square millimeter (kg/mm^2) and fracture toughness of about 6 to 10 $MPam^{1/2}$. PSZ can also provide strengths of up to 100 kpsi at 1500°C—when steel is already molten.[11] Noncubic ZrO_2 particles can be incorporated in other ceramic substances also, e.g., alumina, again with useful results. In this case the best data reported to date are for Al_2O_3 containing 10 to 15 percent ZrO_2, values of fracture toughnesses ranging up to 15 $MPam^{1/2}$, and strengths to 175 kpsi.[13]

Other materials besides ZrO_2 undergo expansive phase transformations, e.g., protoenstatite[14] and Ca_2SiO_4, and it may be that a variety of lighter and less expensive toughening particles than ZrO_2 will be available for exploitation in the future.

The alternative to particle toughening is fiber toughening, and extremely useful progress is beginning to be made in this area, too. Most of the composites prepared to date utilize fine (<10-μm diameter) graphite fibers, or SiC fibers made by the pyrolysis of organic precursors, and silicate glass matrices.[15] Data for fiber-reinforced ceramics are limited at present, but their potential may be surmised from recent data on SiC fiber-reinforced glass ceramics. In this case the matrix was a lithium aluminosilicate glass, and test specimens were prepared with the SiC filament reinforcements present in both unidirectional and cross-ply orientations. In both instances the volume fraction was ~50 percent. Figure 7 presents the fracture toughness data obtained. It ranges from ~17 $MPam^{1/2}$ at room temperature to a remarkable 25 $MPam^{1/2}$ at 1000°C, at which temperature the matrix begins to soften appreciably. For purposes of comparison, the Charpy notch impact strength of this material is about 50 times greater than that of hot pressed Si_3N_4.

In the future we are likely to see the development of ceramics containing several alloying components, processed using complex thermal and mechanical treatments to produce structures equivalent to those in advanced metallic alloys, and exhibiting a dense distribution of crack-arresting, fiberlike entities throughout. A start in this direction has already been made. A number of workers are producing interesting lamellar-type structures by the unidirectional solidification of oxide, carbide, or boride eutectic compositions.[17] To date, their toughness has been somewhat disappointing, typically ~6 $MPam^{1/2}$ or less, but this may be because thc interlamellar interfaces are too strong and so do not produce the desired multitude of crack-retarding microcracks ahead of the propagating major crack. Perhaps the addition of surface-active species that would segregate to and embrittle these interfaces would help.

STRUCTURAL APPLICATIONS FOR HIGH-TECHNOLOGY CERAMICS

Whereas electronic applications for HTCs are already in the marketplace, e.g., alumina substrates and zirconia sensors, products developed to improve on the structural performance of metals in arduous environments are just beginning to be introduced and are still undergoing vigorous technical development.

Certainly the most publicized future application for advanced ceramics is the auto engine.[18-20] However, development of an "all-ceramic" engine of the conventional piston-gasoline variety is not considered likely, both because of design problems and because really substantial gains in fuel efficiency cannot be anticipated. After all, there are metal-engined autos available today providing >50 miles per gallon (mpg).

It seems more likely that over the next few years parts of conventional engines will be produced in ceramics, with different ceramics being chosen for different parts to meet specific operating requirements. Subsequently, new types of engines will be introduced, these being designed

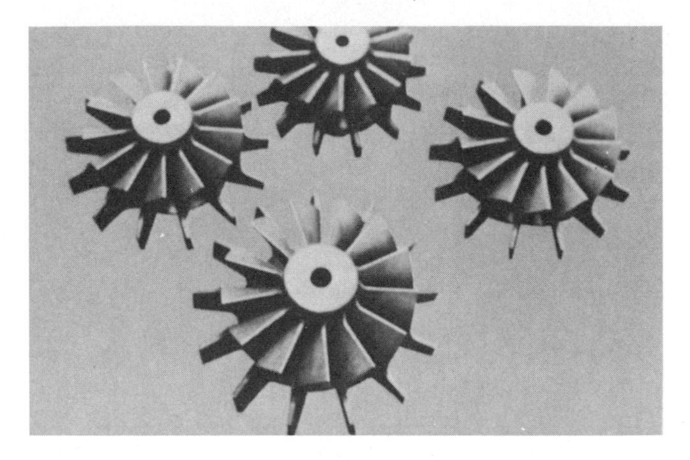

FIGURE 8 Si_3N_4 automobile turbocharger rotors made by Toshiba.[21] Reprinted with permission.

to take specific advantage of the particular attributes of ceramics. The ceramic gas turbine and so-called ceramic adiabatic diesel are the current prime targets.

Over the next five years or so, then, we can expect to see the introduction of ceramic piston caps, cylinder and exhaust manifold liners, valve heads, and turbocharger rotors (see Figure 8),[21] and so forth, because these do not require any substantial redesign of conventional engines, and the control of microstructural flaws is relatively easy. Such components should provide modest improvements in efficiency and durability. Subsequently, with the advent of new types of auto and especially truck engines in the 1990s, we can anticipate the advantages of increased fuel economy by virtue of superior combustion efficiency (because of higher operating temperatures); elimination of cooling (no radiators, water pumps, and so forth); probable elimination of oil lubrication; and reduced emissions. In principle, at least, the cost of such engines should be less than that of current gasoline types.

To demonstrate the feasibility of some of these concepts, the U.S. Army Tank Command (TACOM) and Cummins Engine Company recently teamed up to build an uncooled diesel engine for an army truck (see Figure 9).[22] This truck has operated well so far, providing 9 mpg as compared with the standard 6 to 7 mpg. By mid-1984, the team hopes to demonstrate an oilless version of this engine, with an intrinsic thermal efficiency of ~54 percent as compared with values in the low 30s for conventional gasoline engines.

The ceramics now being developed for auto engine applications include SiC, Si_3N_4 (for turborotors, valves, piston caps, and so forth), PSZ (for combustion-chamber components), and aluminum silicate (for regenerator cores). However, the auto engines of the mid-1990s will probably utilize more sophisticated and complex ceramics than these, most likely alloys with toughness and durability optimized by precise thermomechanical treatments and with surfaces processed ("glazed") to minimize the potential consequences of small flaws introduced by abrasion or erosion. Companies known to be active in the development of ceramic-containing auto engines include GM, Ford, Cummins, Garrett, Volkswagen, Toyota, Isuzu, Nissan, Saab, and Rolls-Royce.

Another area of application is cutting tools. Tungsten carbide was introduced to this application in the 1930s; cemented TiC tools came next, followed by Al_2O_3 in the 1950s, and industrial diamond and cubic boron nitrogen (BN) in the 1960s, the objective always being improved tool life and enhanced rates of metal removal. Recently, however, and as a spin-off from their research on ceramics for gas turbine engines, Ford Motor Company has demonstrated that Si_3N_4 has excellent cutting characteristics, doubling the productivity of conventional cutting tools in the machining of cast-iron auto parts, e.g., wheel drums and clutch components. The Ford material (S-8) contains 8 percent Y_2O_3 and is hot pressed.[23]

GTE has disclosed similar data for its Quantum 5000 Si_3N_4-based material, which contains Y_2O_3, Al_2O_3, and 30 percent TiC as a dispersed phase.[23] They have found that the number of brake drums that can be

FIGURE 9 Army truck used in Cummins-TACOM tests of uncooled, ceramic diesel engine.[22] Reprinted with permission.

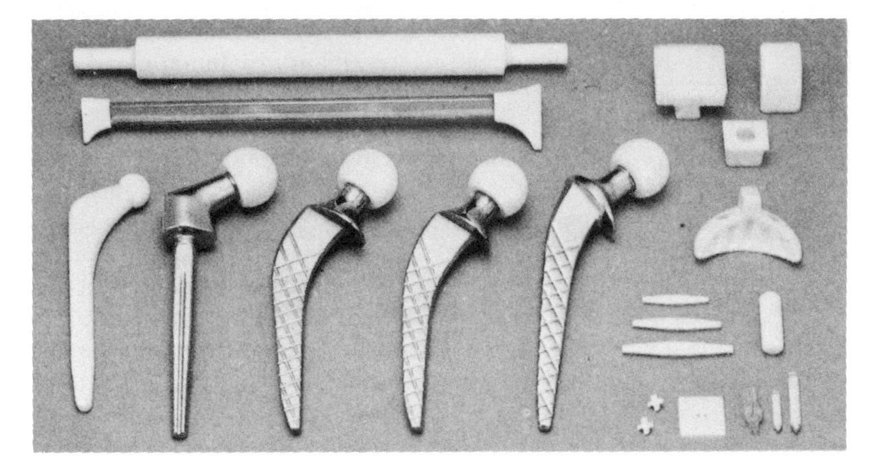

FIGURE 10 A variety of prostheses made from alumina by Kyocera.[24] Reprinted with permission.

produced per tool at cutting rates of ~25 meters per second (m/s) increased from ~10 per conventional tool to ≥100 for the new Si_3N_4-based tools. Other companies, especially in the United Kingdom, are finding that the Sialons (Si_3N_4-Al_2O_3 alloys) also provide excellent tool performance.

Japanese companies, e.g., Toray, are now using Al_2O_3 and PSZ to produce household scissors, nonmagnetic blades for slitting videotape, and surgical saws. They have found that these materials are durable and capable of retaining a very sharp edge.

Other emerging applications for structural ceramics include (1) components required to resist abrasion, erosion, and corrosion, e.g., seals, valves, nozzles, and bearings for the chemical, petrochemical, and mineral-processing industries; (2) armor, especially lightweight body armor utilizing BN backed with Kevlar; and (3) orthopedics. Because of their resistance to corrosion by body fluids and their capability of being formed with surface characteristics closely simulating those of natural bone, ceramics are finding increasing use in surgical applications. Among the many interesting developments in this area are Kyocera's use of single-crystal sapphire to produce a range of products from hip prostheses to dental implants (see Figure 10),[24] and Corning's introduction of potassium-magnesium-silicate ceramic crowns that are cemented directly to the remaining tooth structure and do not require any metal bridgework.[3]

Of course, before high-technology ceramics can truly be considered as "structural materials" and not merely as components of structures, the technology must be developed to join them into integral and self-reliant systems. Work on this problem is underway, and studies at Stanford Research Institute have shown that it is possible to "braze" Si_3N_4 using silicon oxynitride glass compositions similar to those found in the grain boundary phases of this material. Alumina-based ceramic adhesives also are beginning to appear.[3]

Many innovative developments are expected to have occurred in the field of high-technology ceramics by the year 2000. By then, multicomponent, self-reinforced ceramic alloys, heat-treated to optimize properties, protected by compressive surface layers that are perhaps applied by ion bombardment or laser glazing (a new approach to a traditional process), and joined by lasers, electron beams, or novel cements will become respected members of the engineer's portfolio of useful structural materials.

NOTES

1. M.I. Finley, *Horizon*, *9*:51, 1967.
2. E.A. Barringer and H.K. Bowen, *J. Am. Ceram. Soc.*, *65*:C-199, 1982.
3. Reported by R.D. McIntyre, *Mater. Eng.*, p. 19, June 1983.
4. E. Matijevic, *Acc. Chem. Res.*, *14*:22, 1981.
5. E.P. Bond, reported in P. Kennedy and B. North, *Proc. Brit. Ceram. Soc.*, No. 33:1, May 1983.
6. D.C. Nagle, L. Struble, and K. Bridger, Martin Marietta Laboratories, Baltimore, Md. Unpublished work, 1982.
7. Reported by J.R. Hartley, *Auto. Ind.*, *162*:56, 1982.
8. R.A. Graham et al., *Shock Activated Sintering*, Rep. No. SAND-82-2335C, Sandia National Laboratories, Albuquerque, N.Mex., 1982.
9. V.D. Linse and J.H. Adair, *Proc. A.P.I. Conf. on Shock Waves in Condensed Matter*, Santa Fe, N.Mex., p. 14, 1983.
10. Courtesy Powder Technology Center, Division of Kelsey Hayes Corporation, Traverse City, Mich.
11. R.W. Rice, *Chemtech*, p. 230, April 1983.
12. D.L. Porter and A.H. Heuer, *J. Am. Ceram. Soc.*, *60*:183, 1977; *62*:298, 1979.
13. N. Claussen, *J. Am. Ceram. Soc.*, *59*:49, 1976.
14. D.A. Anderson, Martin Marietta Laboratories, Baltimore, Md. Unpublished work, 1982.
15. R.A.J. Sambell, D.H. Bowen, and D.C. Phillips, *J. Mater. Sci.*, *7*:663, 1972.
16. J.J. Brennan and K.M. Prewo, *J. Mater. Sci.*, *17*:2371, 1982.
17. V.S. Stubican and R.C. Bradt, *Annu. Rev. Mater. Sci.*, *11*:267, 1981.
18. J.W. Dizard, *Fortune*, p. 76, July 25, 1983.
19. A.F. Mclean, *Ceram. Bull.*, *61*:861, 1982.
20. D.J. Godfrey, *Mater. Des.*, *4*:759, 1983.
21. "Technical Information—Si_3N_4 Applied Examples," No. 7, Toshiba Corp., Tokyo, Japan, 1983.

22. Reported by S. Robb, *Ceram. Bull.*, *62*:756, 1983.
23. Reported by S. Robb, *Ceram. Bull.*, *62*:206, 1983.
24. *Challenging the Future*, Publication No. Z-101E-1, Kyocera Corp., Kyoto, Japan, p. 20, 1982.

ACKNOWLEDGMENTS It is a pleasure to acknowledge the contributions to this paper made by Roy W. Rice, Naval Research Laboratory, Washington, D.C.; H. Kent Bowen, Massachusetts Institute of Technology, Cambridge, Massachusetts; and a number of the authors' colleagues at Martin Marietta Laboratories, Baltimore, Maryland.

Part IV

Transportation Technology

Introduction

MILTON PIKARSKY

The political, economic, and social consequences of the Organization of Petroleum Exporting Countries' (OPEC's) actions during the late 1973-1974 period, exacerbated by the Iranian embargo in 1979, are still reverberating, and the world's transportation systems—including that of the United States—are in a major transitional period.

The U.S. transportation system can be briefly described as a vast enterprise, the costs of which account for about $500 billion, or in excess of 20 percent of the gross national product, smaller than the amount for housing and larger than the amount for food. It is a fragmented system built and operated by thousands of private firms and government agencies in a decentralized fashion. Relatively few decisions are made, and few individuals or entities have effective control or even influence over much of the system. Rather, the relative size and efficiencies of different subsystems are the product of thousands of lower-level actions that respond to market decisions of shippers and travelers. All of this, when aggregated, results in the total transportation system.

The current technological systems supporting this vast transportation network reflect nearly 100 years of progress, much of which has occurred in the last 50 years. Today, increasingly, technological innovation must be responsive to social, economic, and political pressures if it is to be effective.

The three papers that follow explore current technological developments and trends in these areas of transportation: (1) aviation and aircraft technology, (2) use of reinforced earth in surface transportation projects, and (3) railroad technology.

The U.S. civil aviation industry has been the dominant technological and market force. However, competitive forces worldwide, the domestic economy, and market uncertainties may alter the shape of the industry in the coming years. Technologically the civil aircraft vehicle is highly efficient because of its advanced electronic systems, and increased technological efficiency is possible. However, affordability and planning imply restraint. It is against this background that the industry must shape itself in the future, and in the first paper, John E. Steiner describes the current technology as well as the technological building blocks and the potential rewards and challenges ahead.

In the next paper, James K. Mitchell discusses the cutting edge of technology with regard to soil, one of our most abundant and least expensive potential construction materials. Human attempts to improve the soil for use as a structural material precede recorded history. Indeed, they can be dated to engineering and construction projects of 5,000 years ago, when the pyramids of the shenzi on the Tibetan-Mongolian plateau were built of a compacted mixture of clay and lime.

We are all familiar with the Roman-built roads, which outlasted the empire and are still in evidence today. Soils have been wetted, dried, heated, frozen, pounded upon, vibrated, mixed with other materials, pushed, rolled, implanted with things, and dehydrated to improve their properties. Principles of soil behavior and foundation engineering that are still valid today were recorded in the Tribiuses, 10 books of architecture of the first century B.C. In the New World the Mayans constructed roads that had a base of broken limestone with stone sizes decreasing upward, covered with a mortar of lime and sifted earth. The scarcity of timber in the southwestern United States and regions of Central and South America led to exploitation of local stones and soils, resulting in the familiar adobe brick, strengthened as necessary with sticks and reeds. The quest to better adapt soils for use in construction is not surprising, since in addition to being one of our most abundant and least expensive materials, soil is also one of our potentially most useful structural materials.

Despite the known potential to facilitate earth work and soil improvement work on a large scale, little exploitation of this potential construction material occurred until the twentieth century. Indeed only in the last two decades have the methodologies for soil improvement and ground strengthening been developed. Of the existing methodologies, soil reinforcement is the one most intensely studied and advanced in application. Many of the currently available techniques were specifically developed for use in highway construction. Dr. Mitchell's presentation considers the currently available types of soil reinforcement, their application, design, construction, and economies.

In the third paper, William J. Harris, Jr., discusses technological opportunities in the railroad industry. Successes stemming from earlier technological developments led the industry to a preeminent competitive position. It is natural that under those circumstances the railroads have failed to pursue new technology. However, with the gradual development of the truck and waterway modes of transportation and the completion of the interstate system of highways, the railroads were suddenly confronted by a vigorous economic challenge.

The railroad industry has responded with the adoption of new technologies to force safety improvements and efficiencies in rail transportation. Dr. Harris describes the current status of technology and the anticipated trends in the industry, bringing to the discussion his unique awareness of the technological and research issues facing the railroads during the present, uncertain, transition period on which deregulation legislation has had an impact.

Air Technology: The Transport Vehicle and Its Development Environment

JOHN E. STEINER

Virtually every commercial or military airplane operational today could be technologically superseded by the end of this century. World competition is a forcing factor, but affordability and planning imply restraints. The latest generation of civil transports reflects a significant incremental step into twenty-first century technology. As the technological building blocks offering new efficiencies are validated, the integration task and dependency upon it will increase. In addition, air vehicles must become integrated into a new, advanced-technology, national and international airspace system to attain the important efficiency and safety advantages of total four-dimensional (4-D) strategic control.

TECHNOLOGY AND MARKET NEEDS

The spiraling price of fuel subsequent to the 1973 oil embargo was one of several major influences that had an impact on the direction of jet transport developments during the past decade. The price of fuel as an element of direct operating cost for the trunk airlines is shown in Figure 1. Economic distortions of this magnitude, of course, put much greater priority on the readiness of advancements, contributing to significantly higher orders of fuel efficiency. This signaled a fundamental change to both development and operational objectives, which until that time had been largely oriented toward performance. This is not to say that earlier jet transport developments had not produced efficiency advancements. Fuel efficiency has improved at a very steady rate of about

30 percent per decade over the jet transport's 30-year development history, as illustrated in Figure 2.

The trend line in Figure 2 encompasses a broad spectrum of airplane sizes, and thus scale has little to do with its continued slope. Fuel cost was not the issue that established this trend; however, fuel efficiency has always been highly significant for competitive range performance gains. It is expected that the efficiency trend will continue and, more likely, will increase as the shaded area of the illustration indicates.

More frequently than not, a technological gain will be countered by any number of offsetting factors in its environment—for example, the delay trends at major airports (see Figure 3).

This problem actually preceded the energy crisis by nearly a decade as air transportation's growth accelerated so very rapidly in the 1960s and 1970s. Rapid growth creates its own constraints, as was the case with the imposition of noise regulations and other environmental constraints. However, in terms of today's problems, delays have been greatly exacerbated by airline and hub interchange patterns that evolved with the deregulation of airline competition. Today, with more and smaller aircraft operating from an expanded system of hub interchanges, traffic delays account for as much as a 50 percent nonproductive fuel burn on some shorter route segments. The technology that will virtually eliminate traffic delays is in hand. This vital aspect of aviation technology is discussed later. For now, however, it is well to emphasize that deregulation itself is an indirect contributor to technological changes. Nonetheless, it has become a very powerful contributing influence on the general direction in which jet transport technologies are headed, and, for that matter, on the direction in which the U.S. industry may be headed.

The cost of labor is one of the major competitive problems for pre-deregulation trunk carriers, as Figure 4 illustrates. It is readily apparent from this chart that airline salaries have increased substantially faster than has inflation or the revenue yielded from air fare structures. There is ample evidence that low-cost air fare competition is the most substantial force driving the trunks toward lower cost and more efficient and productive operations.

In like manner these needs have shaped vehicle development objectives—i.e., major improvement in operating efficiency, reduced crew workloads, and growth to 4-D navigation—launching the most recent aircraft types now in service. In this sense deregulation has reinforced operating efficiences as the principal development objective. Development of the Flight Management System (FMS) was accelerated in recognition of airline efficiency needs. However, the readiness of contributing component and systems technology preceded the Deregulation

Act by more than a decade. In fact, significant aspects of FMS's development are rooted in U.S. supersonic transport (SST) work of the 1960s. The FMS evolved from the SST and other independent programs, which were generally oriented toward automation of flight management and control functions. The relationship of these efforts is show in Figure 5.

The FMS is a fully integrated digital electronic system that provides previously unavailable performance optimization and flight management capabilities. The automation and integration of flight control and performance management permit a substantial improvement in direct operating costs, primarily by reducing fuel burn and also by reducing the cockpit crew requirement. This is a very significant advancement with respect to today's efficiency needs, but more importantly the FMS technology represents a vital step toward twenty-first century potentials. Some other evolutionary improvements can be expected to appear over the rest of the 1980s, but most of the new products to be offered in this decade are already known and will be competing for the world open-lift market (see Figure 6).

Several facts are crucial to future U.S. developments with respect to this forecast. First, there is a 40 to 60 percent split in the open market that emphasizes a huge expansion cycle by foreign airlines. Developing nations are expected to form the higher growth segments in this expansion. The second point may be more critical. The earlier timing of foreign market growth will significantly stimulate foreign industry in readying advancements that will be applied to designs for the next generation of jet transports. The U.S. industry could become quite vulnerable in this respect, since an advantageous momentum in developing the visible potentials could technically supersede today's products by the end of this century.

DESIGN BUILDING BLOCKS

The design advancements for twenty-first century transports are embodied in a very large number of potentials that are quite visible to all of the world's aircraft builders. The development of these advancements will doubtless bring vast changes to aeronautical reality as it is known today. Nonetheless, the potentials are so numerous that selectivity among the development options or combinations thereof is itself a problem, and fairly complex, in that the twenty-first century potential is founded on integration complexities of far greater magnitude than those experienced in combining wing sweep with the axial-flow compressor some four decades ago. The integrative aspect is discussed in more detail later, but it should be emphasized here that integration considerations become very apparent at a much earlier stage of development than they did in the past.

The development possibilities among materials illustrate the selection problem. The next generation of transports may well spring from a body of composite technology that has progressed sufficiently to allow composite application to most of the primary structure. We expect this to be the case, with composites accounting for about 65 percent of distributed airframe weight. In this event aluminum use would diminish to about 11 percent (compared with 80 percent in the Boeing 767). However, advanced aluminum developments indicate that another scenario could develop. The options are shown in Figure 7.

The aluminum alternative (right side of Figure 7) suggests that this material's use could amount to over half of the total distributed weight, dropping the projected composite applications to about 25 percent. This alternative scenario developed from the promising advancements in aluminum-lithium alloys, which indicate that aluminum density can be reduced some 3 percent by using a 1 percent lithium addition. The significance of this development is perhaps better understood by recognizing that the empty weight of the 747 could be reduced by about 11,000 pounds through the substitution of this material. Some other promising materials developments currently receiving attention are shown in Figure 8.

In this figure the general materials development boundaries are identified as we presently understand them. The integrative potentials of these or of other materials systems or hybrids with other advancement areas will shape the ultimate design selections. From a fuel efficiency standpoint, however, as much as one-third of the expected improvement potential may be derived from new materials.

Generally the advancement pattern of aircraft development has featured some basis for advancement in a current development that forms a launching pad for a new generation of technology. The all-digital FMS implies this quality because of the multifunction integrations involved. Its development has unquestionably facilitated our ability to capitalize on other avionics-related potentials. Full-scale active control is a natural follow-on. The "full-scale" is emphasized to distinguish future developments from the lesser active control developments known today.

Figure 9 depicts a possible future active control configuration in comparison with today's baseline. The wing is moved forward; the normal center of gravity moves aft; and, as noted, the horizontal stabilizer is significantly reduced in size. At cruise it will carry little or no load as compared with the large downloading on current designs. The full-scale active control development could produce a 5 to 10 percent improvement in efficiency. However, the introduction of artificial flight stability would necessarily emphasize the significance of electronics reliability and systems redundancy to a level well beyond the sizable dictates in today's technology. The air vehicle is rapidly assuming an "electronic system" orientation, and virtually all of the advancement potentials (including the materials advances just discussed) have similar fault-protection concerns.

As the air vehicle assumes a fundamental change in orientation, it can embody some new design concepts that might enhance specific air transport jobs, which will present an exciting possibility for designers. An example would be civil adaptions of the variable camber wing concept shown in Figure 10. Wing flaps and slats have been used for years in securing variable camber and wing cord extension. Most applications have involved drag-producing external structures and considerable weight. There may be some commercial transport applications in which cord extension is not as critical. In such cases an internal hydromechanical, computer-operated, low-drag system of camber variations as illustrated in Figure 10 may prove attractive.

The possibilities of mechanically induced laminar flow-control (LFC) concepts, such as illustrated in the upper portion of Figure 11, have been known and demonstrated in test configurations for many years. Practical solutions for the problems associated with mechanical LFC fabrication and maintenance have proven very elusive, and until recently the concept had retreated from the forefront of aeronautical thinking. However, since LFC answers could produce an exceedingly significant 20 to 30 percent efficiency gain, world interest in advancing both mechanical solutions and natural-flow improvements is again rising.

As with the variable camber wing or some of the other possible civil transport potentials, a "second look" at earlier concepts is premised on the synergy of a totally new environment of advanced electronics and electrics, the evolutionary aspect of which is shown in Figure 12. The development trend is leading toward a completely different concept in flight control for civil transports and also toward vastly differing relationships between the flight crew, the air vehicle, and the air system in which both operate. The all-electric developments imply a substantial reduction and possibly a future total elimination of hydraulics and control cables, resulting in significant weight savings. It should be remembered that weight reductions translate into a smaller vehicle for a given job, thus reducing the cost of manufacture and acquistion. Elimination of cables also means that the familiar cockpit yoke will disappear, bringing different geometry considerations into the overall cockpit design.

Electronics-based advancements are key in attaining future air transportation potentials, and they have evidenced a growing significance in vehicle improvements made thus far. Rapid growth in avionics developments is, of course, directly related to the digital avionics advancements discussed earlier. Many of the potentials illustrated in Figure 12 are in advanced development stages now. The insertion of digital processors into sensory and control systems is also one of the more current refinements that has been utilized in propulsion system developments. Commercial jet engine advancements have produced a 40 percent improvement in specific fuel consumption over the last two decades, with more of the efficiency thus far centered on high bypass-ratio (BPR) engine developments. These trends are illustrated in Figure 13.

While engine efficiency improvements are expected to continue at about 20 percent per decade, future engines may not much resemble those familiar today. Steady improvements in design and materials appear to have made future gains less dependent on bypass ratios. The BPR increase trend in fact has a negative side in that it increases nacelle weight and, of course, contributes to a mismatch between takeoff and cruise thrust requirements. This does not mean that the high-bypass turbojet has reached its improvement capacity. It does, however, indicate that progress toward an unducted fan or turboprop is starting to look more attractive. Once design considerations move toward a geared cycle engine, the number of possible alternatives increases significantly. We should expect that some form of propellerlike machine (a propfan or new-generation turboprop) will attain operational status before the end of this century.

The many possibilities in the propulsion area amply demonstrate that the potentials in aeronautics will clearly outweigh the resources available for development. One of the most difficult tasks over the next few years will be to make the right selections and to define an orderly development plan. Designer attention is always focused on the critical mass of technology that is available. The building blocks discussed above are forming a new critical mass for future design focus. The relationship with present-day technology is shown in Figure 14.

The building block "aiming points" (right side of Figure 14) are shown in relationship to a baseline (lower left) representing the 1970 efficiency level achieved by the 727-200 technology. The "improved product efficiency levels" indicate the 1980 decade's technology gains that are incorporated into the latest new aircraft types. At each level, the 1970 baseline and beyond, a critical mass of available new technology was ready, and based in each were distinctive advancement threads leading directly into the generation ahead.

Much of current industry attention is focused on the smaller-sized transports in the 100- to 180-passenger range. However, it should be noted that airplanes for a given job have become larger over time. It can be expected that the established capacity growth trend will continue. Economics of scale are involved and, of course, so is the predicted growth in revenue passenger and freight ton miles. This is mentioned because each step of increase has involved additional considerations and solutions for passenger accommodations both in the air vehicle and the terminal area (for loading and unloading) and also in airport access and egress. Fundamentally there is no technical limitation to size, but there will be a continued need for technical solutions and infrastructure changes to accommodate increased size. This need is illustrated by the current generation of transport growth potentials shown in Figure 15.

The cross section shown in Figure 15 is that of the 747, which already has a 500-passenger high-density configuration for the Japanese domestic market. The illustration shows that the more conventional passenger payload for this airplane type could nearly triple by expansion to a full double-deck configuration. All airline markets differ, and while many airlines are absorbed in smaller aircraft solutions, others are exploring potentials of the nature illustrated. Obviously airport infrastructure, including the feasibility of double-deck access/egress systems, would have to be considered very seriously in such situations.

READINESS: A NATIONAL PROBLEM

One of the most perplexing problems facing the United States is our ability to exploit these cutting edges of technology in a manner that will most benefit the nation's economic and national security. In the arena of global high-technology competitions, these two objectives are increasingly viewed as being one and the same. The United States is uncomfortable with this view, partly because of our preferred divisions in public and private sector responsibilities. I am convinced, however, that in terms of our national understanding of technology, the greater difficulty lies not so much in areas of substance as in the advancement chain itself, namely, the progression from (1) fundamental knowledge (research) to (2) technology validation (readiness) to (3) product applications (production).

This advancement chain has two critical stages beyond the basic research that produces fundamental knowledge and reveals new technological potentials. These recognitions come from multitudinous efforts and sources: government, academic, and industrial research facilities in many countries. I do not believe that the greatest area of U.S. national discomfort is with this stage, or even with the last, in which technology exploitation occurs. The American industrial system has proven to be very effective here, providing that technological risks have been reduced to acceptable levels. To my mind, the most critical problems are centered at the midphase—technology validation. U.S. advancement is much in jeopardy because we have become a nation completely at odds with the requirements that this stage entails. The requirements of technology validation are as follows:

- Midphase is the most critical, lengthy, and expensive part of the technology process.
- Risks are identified and reduced to product acceptable levels before a production commitment is made.
- "Validated" advancements can frequently be exploited as improvements in current product line.
- A "critical mass" of advancements may justify a new program start.

The major difficulty is that the validation and application readiness of any advancement take a great deal of time and money plus carefully orchestrated continuity of effort (see, for example, the laser gyro system development history shown in Figure 16).

This strap-down development is a key component of the FMS advancement described earlier. Its application readiness became highly significant to the efficiency gains made by the latest generation of transport aircraft. The 20-year program of development and application readiness surrounding this one component was a multithreaded effort that built the body of technology and readiness acceptability on which its production commitment rests. There is no satisfactory way to circumvent the process or the need. However, today many people in national leadership positions wielding tremendous influence over technology do not understand the process or the need, which are shown in the following list.

- Readiness attainment is absolutely vital if production program costs and risks are to be acceptable.
- The task starts with identified potentials (civil view), not with identified mission requirements (military view).
- Producibility is part of the readiness task. Attainment involves the entire manufacturing/supplier base.
- Readiness is the least understood link in the innovative chain.

Because the readiness attainment process is not understood by many in positions of national leadership, we have self-imposed barriers to readiness that diminish the U.S. capacity to innovate. I believe that most of these barriers are unintended constraints, but they are nonetheless real. For example, in aircraft developments today there are requirements for an "audit trail," linking readiness tasks to an identified military mission need—a specific weapon system.

This one requirement places a top-downward constraint on what is logically the reverse—a bottom-upward process of building. Seldom is it fully recognized that failures, delays, or cost overruns in weapons system developments may have been preordained because constraints of this nature short-circuited the midphase. An adequate job requires more time, effort, and money than many in the United States would care to admit. However, readiness attainment is crucial for successful production to occur, and other nations have been willing to pay its price. This is evident from the shifts in aeronautical leadership positions that are displayed in Figure 17.

Despite slowed momentum, the United States still has the outstanding foundation for aeronautical attainment in terms of the breadth of its high technology—in computers, propulsion, electronics, materials, and other areas, which will be united at the cutting edge of future air vehicle technology. However, it must be kept in mind that no matter how impressive this foundation appears today, it is vulnerable to the pace and continuity of foreign advancements.

A NEW AIR ENVIRONMENT

The twenty-first century potentials for aircraft, of course, will not be realized without modernization within the air system as well. In this area there is little doubt that the United States has assumed a leadership position, and for good reason. Today the United States operates the busiest air traffic control system in the world, and it does so with a remarkable level of efficiency and safety, given the workload. The U.S. national airway system today can be summarized as follows:

- Air traffic control and air navigation
  - 233,000 aircraft
  - 3,200 airports under system control (12,800 others uncontrolled)
  - 126 million aircraft operations annually (four each second)
- Mixture of old and new electronic control equipment

The current system has evolved piecemeal over the 40-year expansion in U.S. domestic air travel. As it functions today, it is expensive to operate and maintain and has little room for the expansion that future traffic growth will require. The limited capacity for expansion is particularly sensitive since system saturations at key airports are already a serious factor, as was noted earlier. When combined with operations projections (shown in Figure 18) that indicate present traffic loads will more than double, it became clear that modernization would entail a total, phased restructuring of the system.

Airline carrier domestic enplanements by the year 2000 will increase by more than 120 percent, and, of course, this will place extreme pressures on the 27 key U.S. airports that now handle about 70 percent of all domestic passengers. Also, very few new airports are expected to be built by the turn of the century. The National Airspace System Plan (NASP), initiated in 1982, which will bring U.S. airspace into the twenty-first century, was developed with the following objectives:

- Increase in control capacity (doubled by year 2000)
- Increase in airports controlled (410,000 aircraft operating from 4,000 controlled airports)
- Elimination of serious delays (4-D navigation)
- Improvement of safety
- Reduction of system and user operating costs

It is important to note that the many years of effort that went into the plan's formation involved the entire aviation community, and the detailed phasing of implemention has been carefully coordinated with the airspace user. The fundamental elements of NASP are as follows:

- A national computerized integrated system
  - Traffic control
  - Host computer
  - Solid-state radar
  - Automated data link
  - Integrated national telecommunications network
  - Microwave landing systems
- Approximate cost: $10 billion over 10 years
- Eventual interface with most other world areas

The transition and evolution to a year 2000 system architecture that incorporates all the elements noted above are challenging under any circumstances. However, bringing the new U.S. air system to reality is a task something equivalent to the Apollo program. It is all the more challenging in that phasing and the many integrations it involves must not interfere with the around-the-clock operation of the present traffic control system.

The integrative challenges are enormous, and for this reason there will be an overall systems engineering and integration contractor selected to oversee the total effort. By the year 2000 we can expect to see computers in the air talking to computers on the ground to control navigation. In this respect the revolutionary changes wrought by digital electronics in aircraft systems and flight management are also extended into integration with the air system's ground-based elements in all 50 states, territories, and oceanic regions. The U.S. plan has been endorsed by the governments of Canada and Mexico. Eventually, it is hoped that the American system will find compatible interfaces with other world areas.

INTEGRATION: A NEW DISCIPLINE

If anything, this brief look at the cutting edge of aeronautical technology tells us that the sum of twenty-first-century potentials will be derived from their integrations. Integration tasks will focus on three fundamental areas: (1) the air system, (2) the air vehicle, and (3) the airport environment (access and egress). Much will rest on how well we understand integration and its implications in a conceptual sense, on how thoroughly it is explored in risk validation, and on how efficient and effective the methodologies and processes employed for this are.

Integration might be seen as a new discipline, considering what is not known about it at present and what this may imply for the more traditional aeronautical technologies and processes. There is little doubt that the new digital electronics orientation of the air vehicle has caused us to reconsider design concepts in a completely different manner. The familiar tools and processes for research and analysis—flight testing, simulation, and wind tunnels—may also assume different significance.

Integrative aspects of the cutting edge, for example, could amplify the role of simulation in design, testing, and verification over the coming years. This is not to say that wind tunnels, flight testing, and flight demonstration will become less significant. It does mean, however, that the new vehicle orientation would emphasize digital-based flight simulation techniques as excellent tools for detecting integration risks inherent in many of the new digital-based advancements. Flight simulation is a powerful development tool that will find many significant engineering analysis, development, and test applications as a better understanding of integration is achieved.

Perhaps the first appreciation of this was gained when it became apparent that the subsystem integration necessary for development of the FMS concept also required the integration of the somewhat compartmentalized testing capabilities into a cohesive, ground-based simulation of a flying test-bed environment, far more complex in total than any simulation previously envisioned. Put another way, it was recognized that having "the right stuff" in on-board computer software is as vital for the machine as it is for the human linkage. It was necessary to satisfy ourselves, as well as the machine, that what we were doing was valid and reliable. Figure 19 shows one of the test consoles used for integration and validation of flight management subsystems.

The FMS and the techniques used for its development and verification were significant steps toward completely different relationships between the air vehicle, the flight crew, and the air system environment in which both will function. Some of these changes are apparent in the advanced flight deck concept shown in Figure 20. With the potential of an all-electric airplane, control cables will disappear, and cockpit-space relationships will become radically different. Digitized voice communications can relieve much of the flight crew communications workload, and flat-panel CRT (cathode ray tube) displays will provide integrated flight progress and flight management information to the crew. Integration is one of the keys to putting these very complex elements—man, machine, and air system—into the safe, reliable, and affordable relationships that we envision.

In this context, the integration trends evident in aeronautical advancements are paralleled by integrative trends that affect virtually every aspect of the industrial process: technology, design, production, and support.

Efficiency, productivity, and affordability are all major competitive influences driving both technology and processes toward higher orders of integration. The cost of advancement has very rightfully been challenged by the affordability of advancement. Figure 21 frames this complex problem with respect to the nation's air system. The really crucial questions for civil aviation and its innovative capacities are clustered around this single issue. Technical solutions are vital, but the best technical solution must be as acceptable to the air traveler as it is to the manufacturers and airlines. All will have a role in determining what is affordable.

CONCLUSION

The traditional technologies associated with aeronautics (aerodynamics, structures, materials, propulsion, and systems) all have very sizable development potentials that appear feasible over the coming decades. Integrated, these potentials offer revolutionary levels of advancement for the civil transport vehicle and also for the air system in which it operates. The possibilities for this are founded on higher orders of digital electronics advancement—also the keystone for revolutionary changes that are integrating the processes of advancement and those of production.

The trend of the future is toward integration; the reason is affordability, a problem of both national and international dimensions. It will demand that the compatibility of technologies, processes, and people be thoroughly understood. This is the challenge, and we stand at the cutting edge of its solution.

ACKNOWLEDGMENT The author gratefully acknowledges the contributions of others, especially that of L. K. Montle.

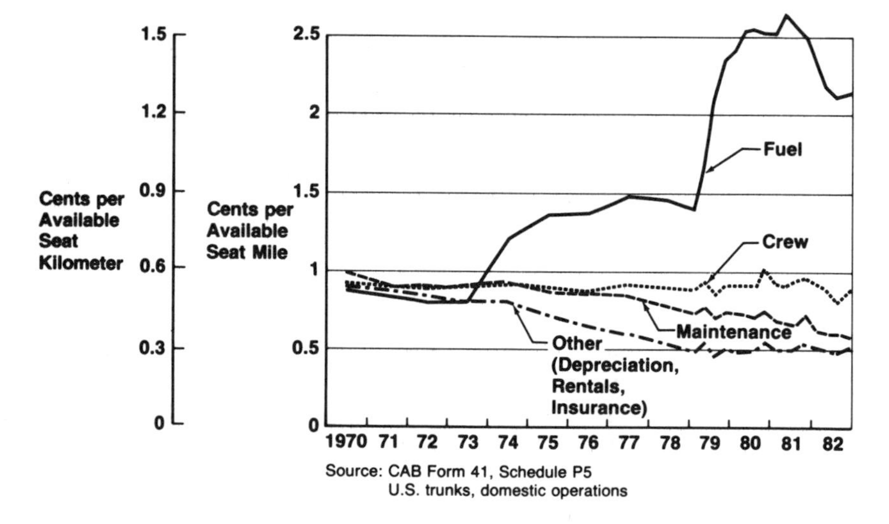

FIGURE 1 Influence of fuel price—direct operating cost elements (constant 1982 dollars).

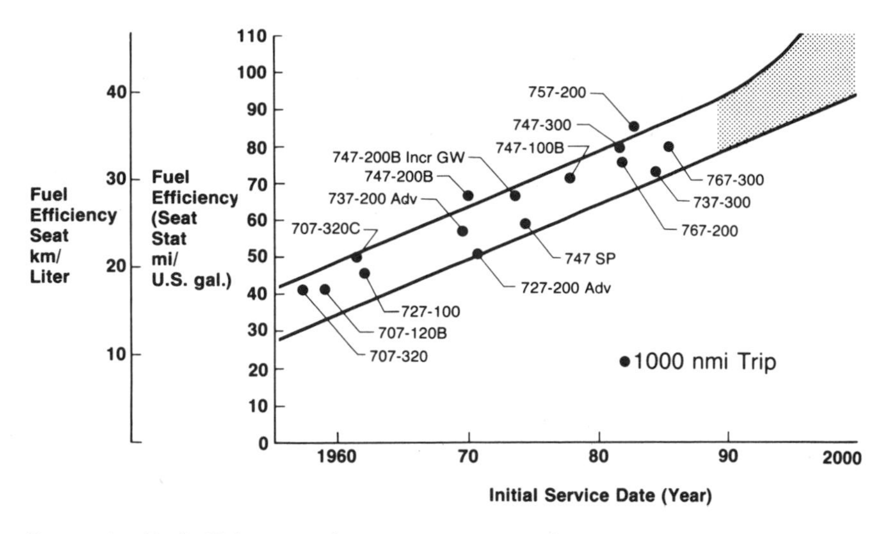

FIGURE 2 Fuel efficiency trends.

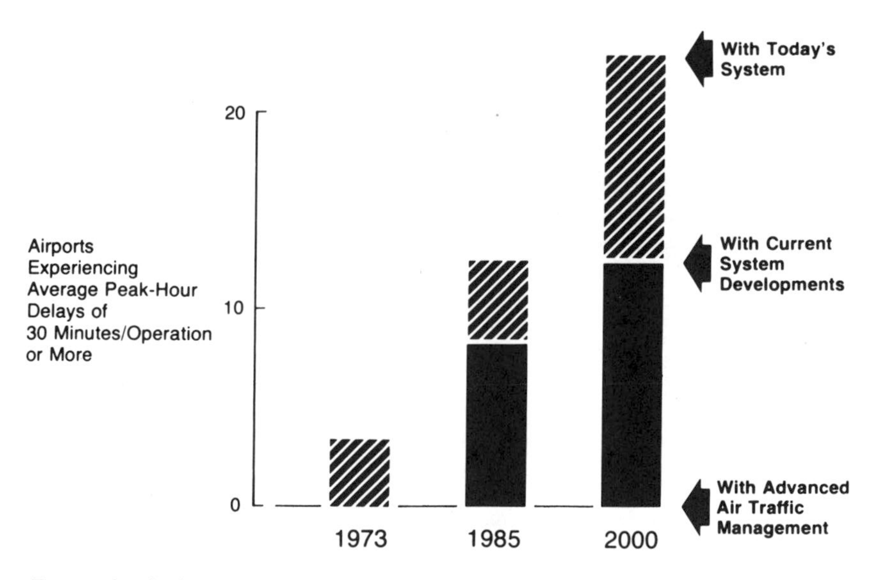

FIGURE 3 Delay trends at 25 major airports.

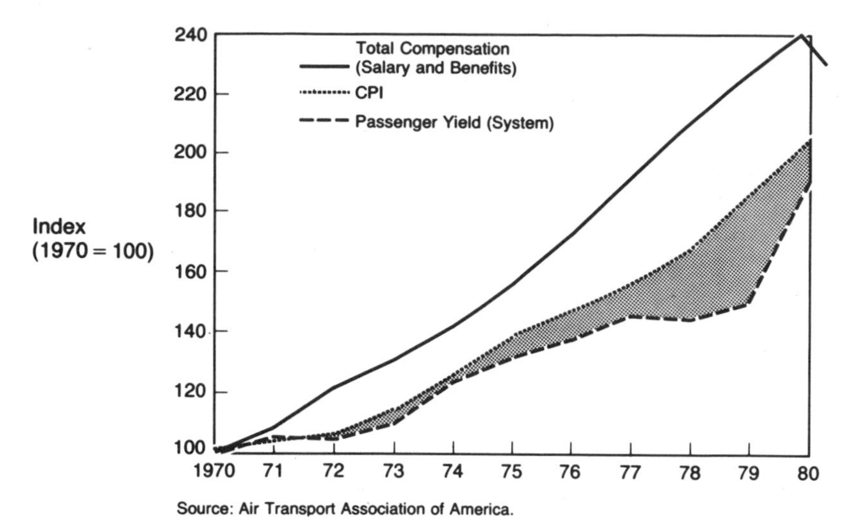

FIGURE 4 Total compensation per airline employee versus consumer price index and yield. Reprinted with permission.

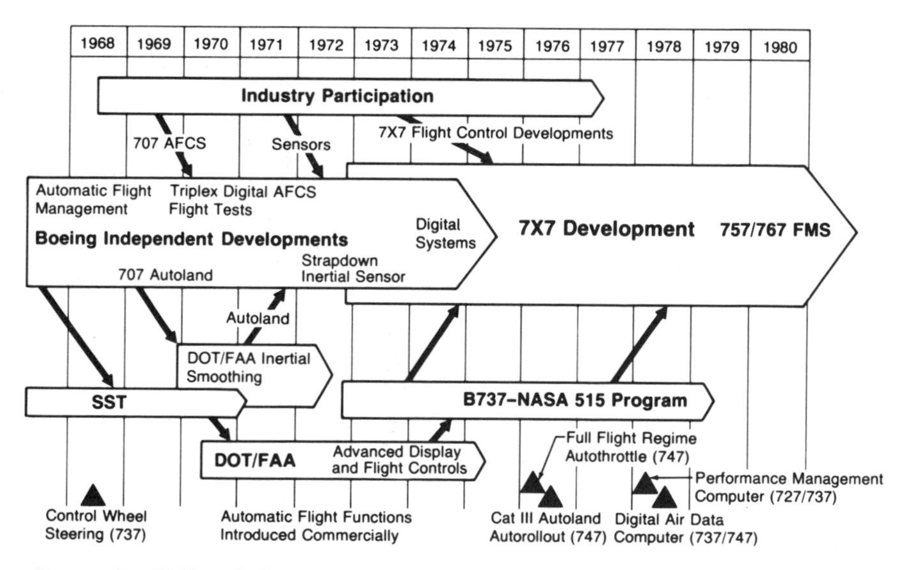

FIGURE 5 FMS evolution.

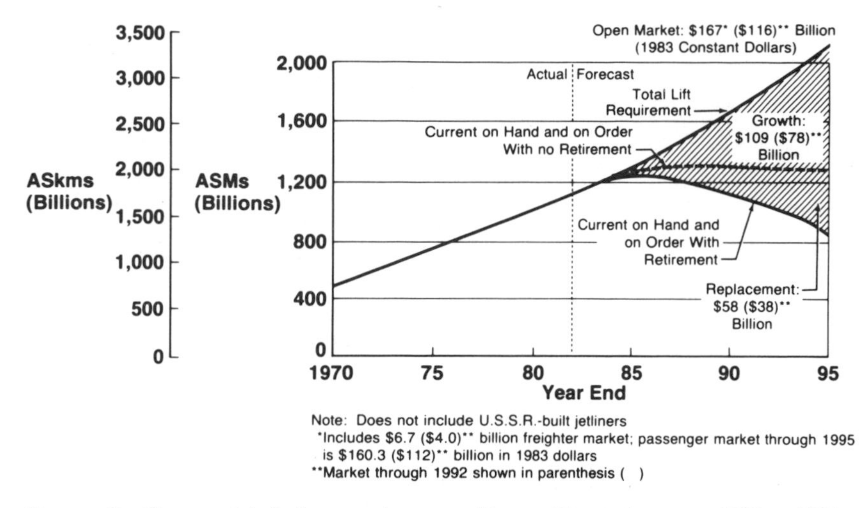

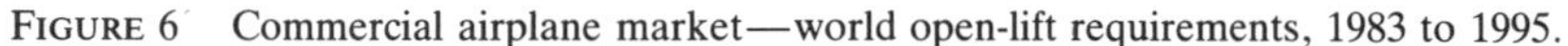

FIGURE 6 Commercial airplane market—world open-lift requirements, 1983 to 1995.

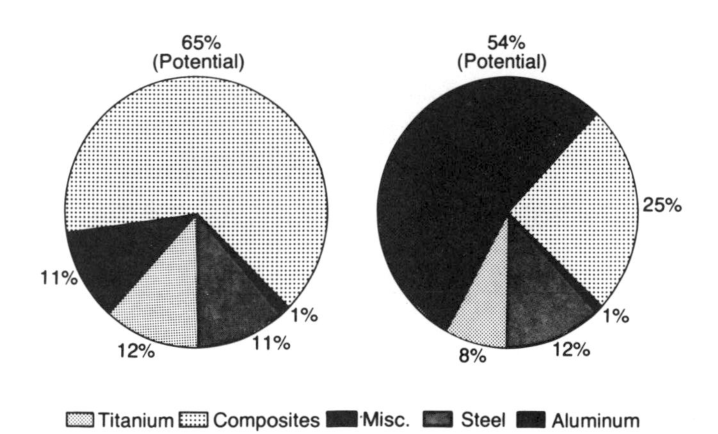

FIGURE 7 Potentials for 1990 subsonic airplane—materials weight distribution.

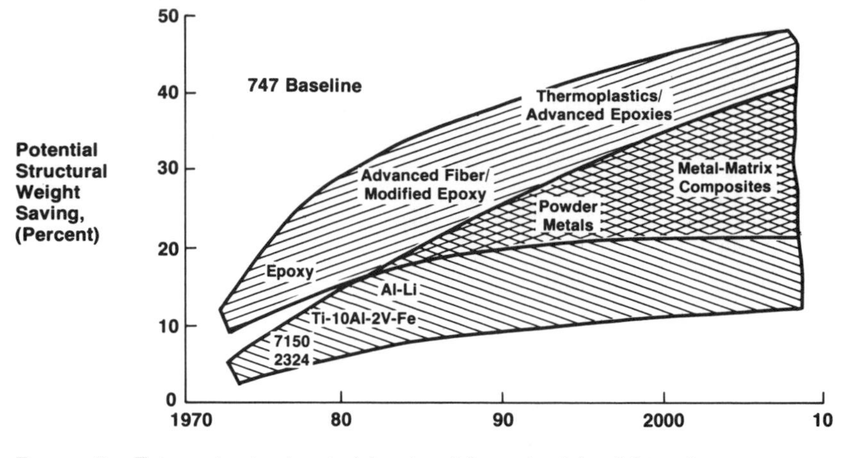

FIGURE 8 Future structural materials—trend for potential weight savings.

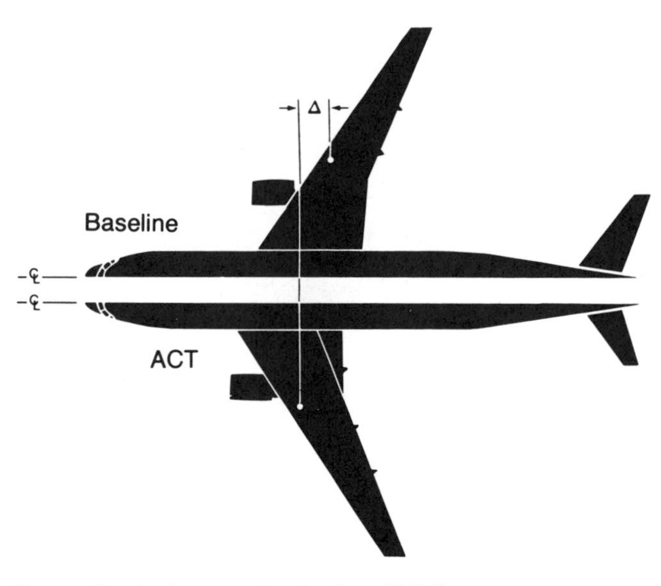

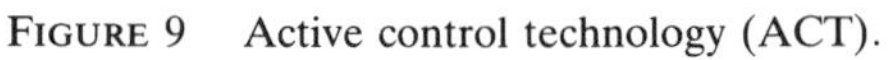
FIGURE 9 Active control technology (ACT).

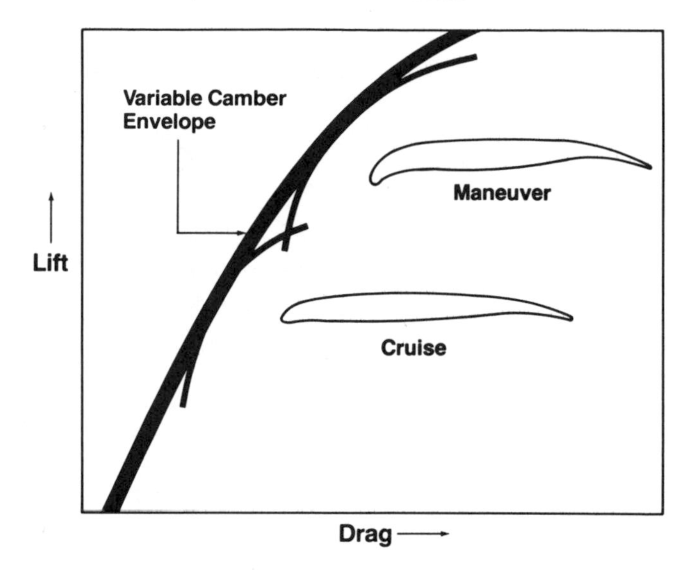

FIGURE 10 Computerized airfoil camber control.

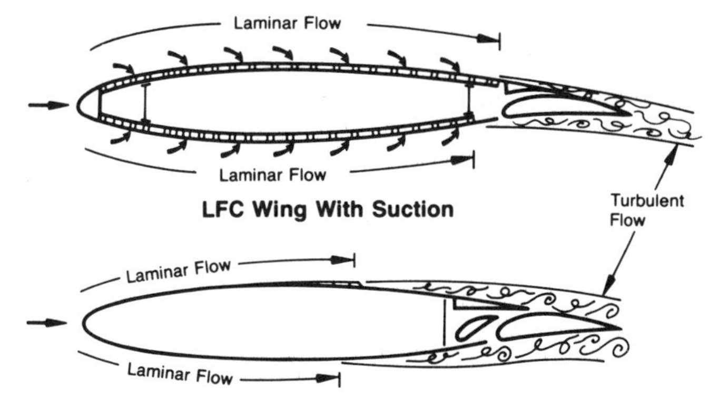

FIGURE 11 Laminar flow.

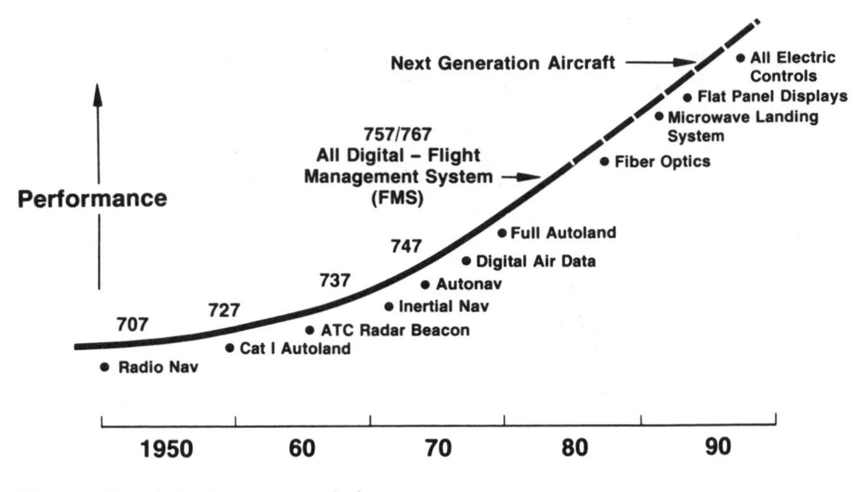

FIGURE 12 Avionic system evolution.

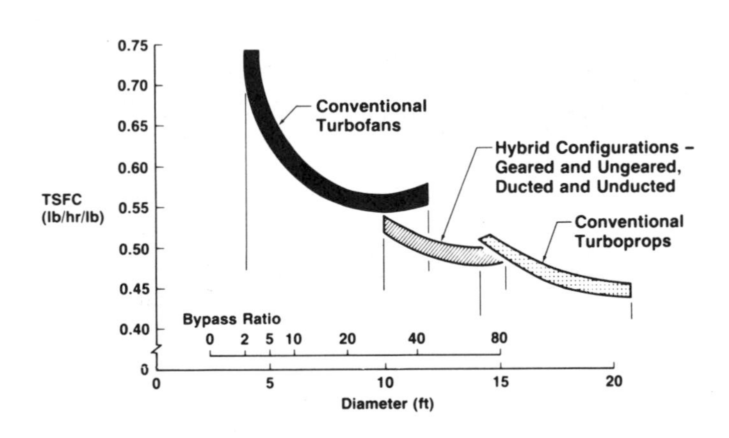

FIGURE 13 Specific fuel consumption—engine size trends.

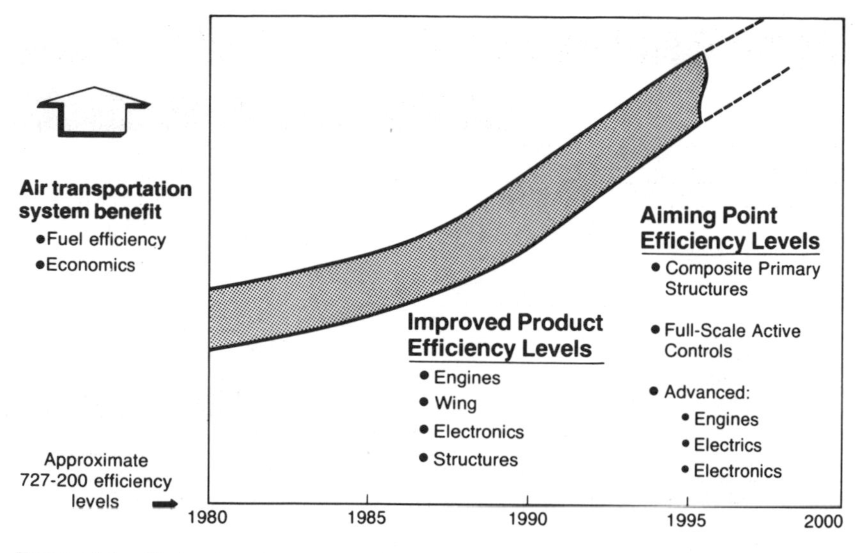

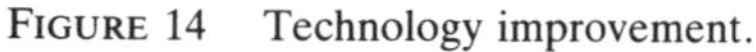
FIGURE 14 Technology improvement.

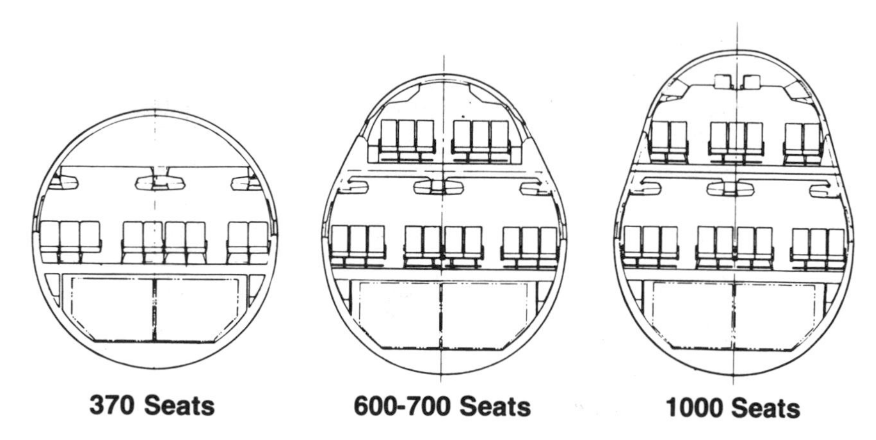

FIGURE 15 Growth potential.

20 Years of Ongoing Laser Technology
12 Years of Application Readiness
Theoretical Development
Laboratory Development and Evaluation
• Processes
• Software
• Producibility
• Testing
Technology Application
Advanced Gyro Development
• Math Models
• Materials
• Block Design
Systems Development and Applications
Production
Flight Testing (14 Aircraft Programs)
62 64 66 68 70 72 74 76 78 80 82

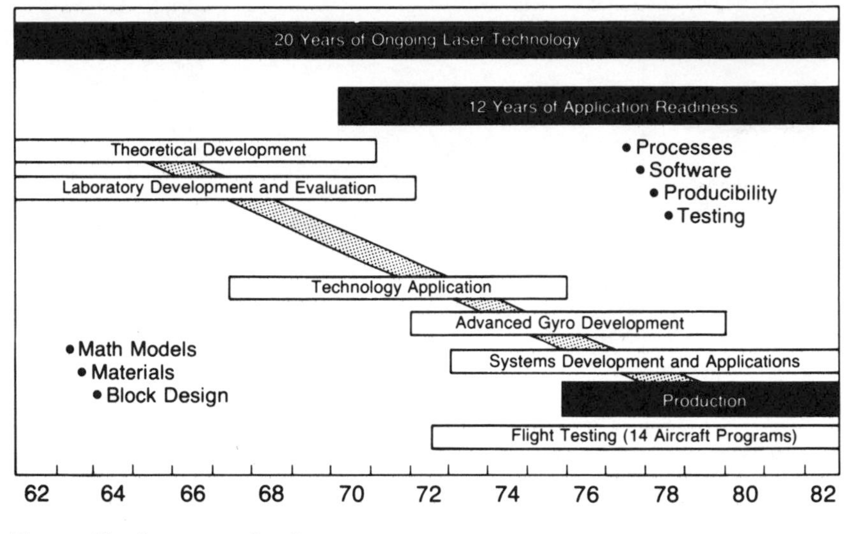

FIGURE 16 Laser gyro development.

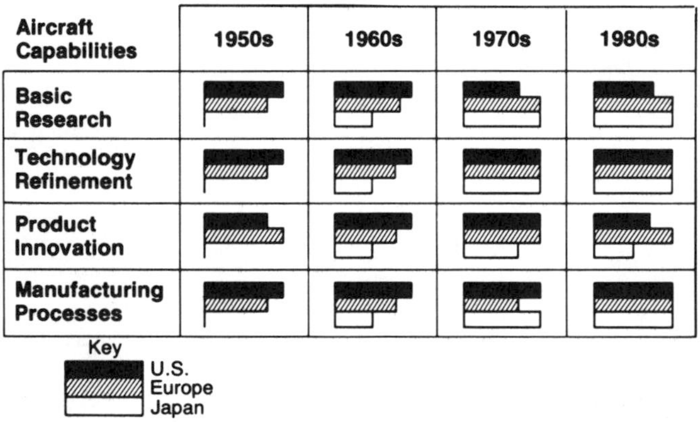

Aircraft Capabilities
1950s
1960s
1970s
1980s
Basic Research
Technology Refinement
Product Innovation
Manufacturing Processes
Key
U.S.
Europe
Japan

FIGURE 17 Shifts in leadership momentum.

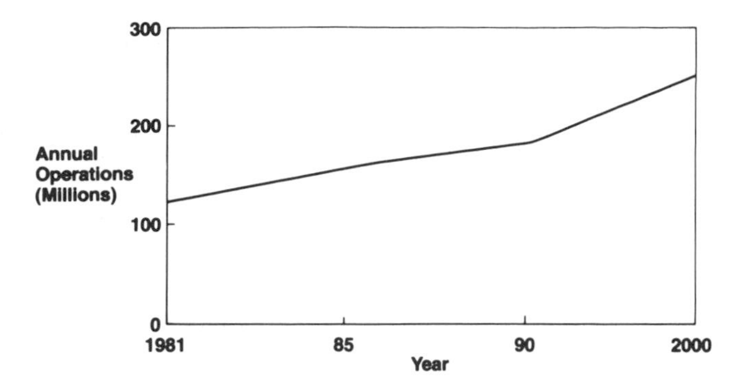

FIGURE 18 U.S. aircraft operations forecast.

FIGURE 19 FMS system integration console.

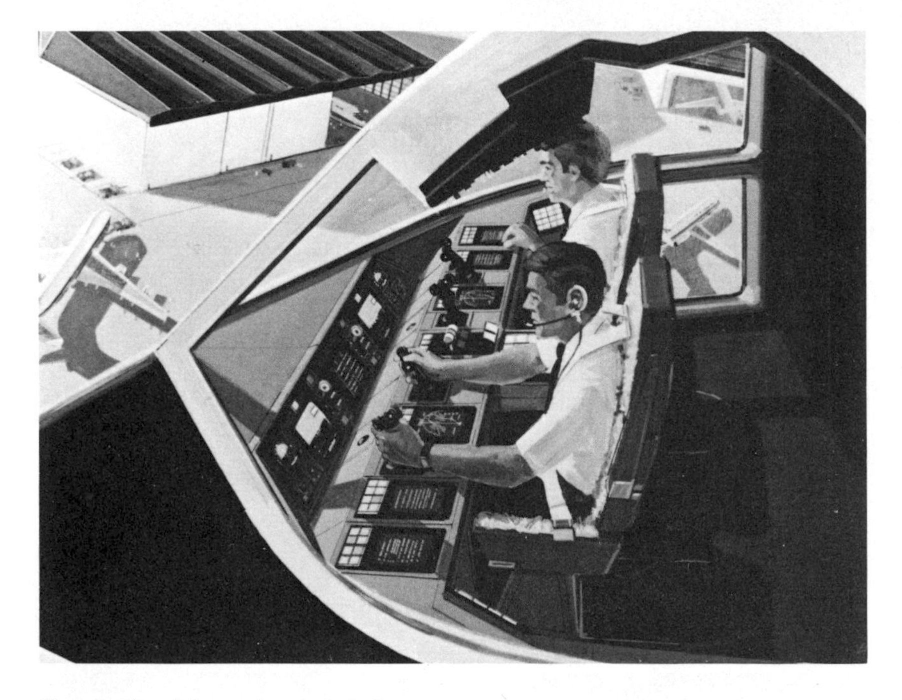

FIGURE 20 Advanced cockpit design.

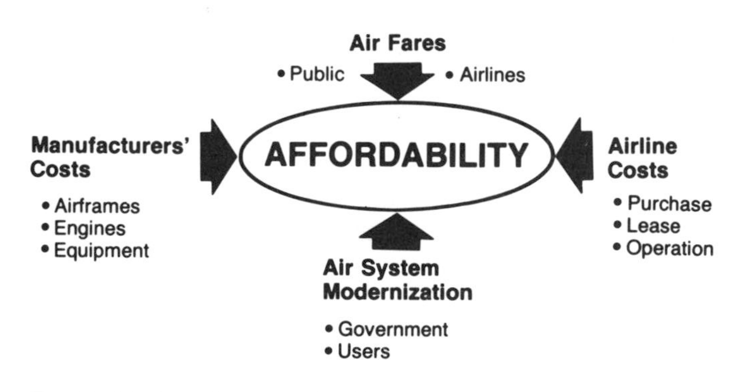

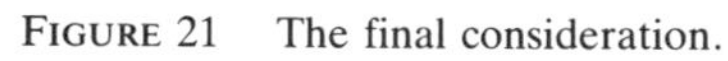

FIGURE 21 The final consideration.

Earth Walls

JAMES K. MITCHELL AND JAMES G. COLLIN

Soil is our most abundant and least expensive construction material. The range of soil types and possible states for any soil type are almost limitless. When at a suitable density and moisture content, most soils can be strong enough in compression and shear to be structurally useful. On the other hand, like portland cement concrete, soil is very weak in tension; this limits its use for some applications, such as those requiring slopes steeper than the internal friction angle of the soil, which is about 30 degrees in most cases. But also, as is the case for reinforced concrete, the inclusion of reinforcement that is strong in tension yields a composite material that combines the best features of both components.

Figure 1 shows a pile of sand with the steepest slope that can be maintained with the sand in its dry state. The same sand with sufficient water added to give capillary stresses in the pores, i.e., fluid pressures less than atmospheric, results in positive contact stresses between sand particles, or an "apparent cohesion" that permits steep slopes to remain stable, as shown in Figure 2, up to some particular height. Because either wetting or drying of the damp sand will lead to collapse of the slope, the use of capillary stresses can hardly be relied upon for long-term stability.

On the other hand, the inclusion of reinforcements within the sand can, as a result of stress transfers between the soil grains and the reinforcements, result in structures that are stable over long time periods. Figure 3 shows the same dry sand as that in Figure 1, but in this case strips of paper are incorporated as reinforcing elements. The paper used as the wall facing is needed to prevent running of the sand from the region between reinforcements; however, it does not assume a major structural or load-carrying function.

Over the past 15 years increasing use has been made of earth reinforced in a manner similar to that shown in Figure 3 to construct new walls or to strengthen existing slopes. There are several reasons for this in addition to the low cost and abundant supply of earth. Construction is simple and rapid. Reinforcements and facing elements can be prefabricated. Aesthetically pleasing structures are possible. Earth walls do not require unyielding foundation support as is the case for most reinforced concrete walls. Keen competition among the developers of different reinforcement systems has led to rapid technological development and continued cost reductions relative to both traditional types of reinforced concrete walls and other types of wall systems such as crib walls and bin walls.

EVOLUTION OF EARTH WALLS

It has been known for several thousand years that tensile inclusions in soil can provide reinforcement. Large religious towers, called ziggurats, were built by the Babylonians between 5,000 and 2,500 years ago. These structures (see Figure 4) had walls faced with clay bricks in an asphalt mortar with blocks of sun-dried mud behind. Layers of reed matting were laid as horizontal reinforcing sheets in the mud. In some ziggurats additional reinforcement was included in the form of ropes about 50 millimeters (mm) in diameter placed perpendicular to the wall and regularly spaced in the horizontal and vertical directions.

Reference is made in the Old Testament (Exodus 5:6-9) to the use of straw-reinforced bricks by the ancient Egyptians. Many primitive peoples used sticks and branches for reinforcement of mud dwellings. The corduroy road was an early method for construction of roads across very weak ground that was widely used in colonial America. Figure 5 shows a more modern example of a corduroy road. During the seventeenth and eighteenth centuries French settlers along the Bay of Fundy used sticks for reinforcement of mud dikes.

The development of earth reinforcement for walls in its modern form was pioneered by the French architect and inventor Henri Vidal. The results of several years' study and experimentation led to Vidal's patent in 1966 for *Terre Armee*. The first highway use of a Vidal reinforced earth wall was near Nice, France (Figure 6). A schematic diagram of this type of wall is shown in Figure 7.

Vidal reinforced earth walls were first used in the United States in 1972 to provide support for California State Highway 39 along a steep slope in the San Gabriel Mountains north of Los Angeles (Figure 8). An earth wall was well suited for the site because of the overall instability of the hillside and because of the ability of these walls to withstand substantial deformations without failure.

CURRENT REINFORCEMENT SYSTEMS FOR EARTH WALLS

Over the past 10 years the Vidal reinforced earth wall has been the most widely used. Galvanized steel reinforcing strips are connected to precast concrete facing panels, which have largely replaced the metal facings used in early walls. The standard precast concrete facing panel is shown on the walls in Figure 9. For aesthetic reasons special finishes or panel designs can be used, as shown in Figure 10.

A number of other reinforcing systems and materials applications have been developed for earth walls as well. In the VSL Retained Earth system, horizontally placed wire or bar mesh systems are used, as shown in Figure 11. The Hilfiker welded wire wall (Figure 12) uses wire mesh reinforcement and a facing consisting of wire mesh covered by shotcrete.

Tensar geogrids are high-strength plastic composite grids that can be used in a variety of ways for containing, retaining, and reinforcing soil. Since the grids can easily be interconnected, a variety of combinations is possible. Large (1 m × 1 m × 1 m) earth- or rock-filled baskets similar to gabions can be stacked on top of each other to form walls or barriers. The plastic grids can be used during reconstruction of failed earth slopes to improve stability. Figure 13 shows an example of a brick-faced retaining wall using tensar geogrids.

Synthetic fabrics for geotechnical use, called geotextiles, have been used as earth reinforcement. They are particularly suited for relatively low walls along remote or relatively lightly traveled roads. An example is shown in Figure 14. A schematic diagram of the internal structure is shown in Figure 15.

The "Kabil Stack-Sack Wall" shown schematically in Figure 16 is also a simply constructed system that is adaptable to remote sites. The facing consists of vertical reinforcing bars over which sacks full of premixed but just wetted concrete are placed by dropping the sack over the top of the bar. Wire mesh or chain link fencing is used for horizontal reinforcement, as shown.

Reinforcing systems are also used to strengthen existing ground in order to improve slope stability or to enable slope steepening on excavations without internal bracing or active anchor systems. Soil nailing (Figure 17) consists of the insertion, by driving or drilling, of reinforcements into a natural or cut slope. Grouting around the reinforcements is often used to assure good bond between the soil and the reinforcements. "Root piles," consisting of small-diameter, cast-in-place concrete piles containing a single reinforcing bar down the center are also used. Root piles are commonly installed at different inclinations, as shown in Figure 18.

APPLICATIONS IN TRANSPORTATION SYSTEMS

Earth reinforcement systems of various types have become extensively used in transportation projects. Probably the two greatest uses are for retaining walls and bridge abutments, where they compete very favorably, economically and aesthetically, with reinforced concrete. A bridge abutment application is shown in Figure 19. They are equally useful for rail networks. The need to rehabilitate and expand existing road and rail systems in urban areas portends increased use in rights-of-way presently containing earth slopes that will have to be steepened or removed in order to provide the needed space. Any situation requiring an elevation change of more than a few feet is potentially suitable for use of an earth wall.

Earth walls have also been used for waterfront structures, e.g., quay walls. Here the facing elements play an additional role, namely, erosion protection for the soil behind them.

DESIGN CONSIDERATIONS

Earth walls are subject to the same external design criteria as those for a conventional retaining wall. That is, they must be stable against sliding due to the lateral pressure of the soil retained by the wall, they must resist overturning, and there must be safety against foundation failure. Classical methods of soil mechanics have been used for the analyses necessary for this part of the design, and they have been satisfactory.

The internal design of the wall itself must ensure against (1) failure of reinforcements in tension, (2) pullout of reinforcements, and (3) loss of reinforcements by corrosion or other forms of deterioration.

To ensure against the first two failure modes requires knowledge of soil-reinforcement interactions. These depend, in turn, upon soil type, reinforcement type and geometry, and the stress state of the soil. The effective friction coefficient between the soil and reinforcements controls the stress transfer between the materials.

Many analytical, numerical, model, and full-scale field experiments have been done to provide information on which the selection of needed parameters for design can be based. Sufficient information has been obtained from these studies and from experience with existing walls that safe designs are possible for normal loading conditions.

The soil used in the reinforced zone of permanent reinforced earth walls is required to be cohesionless so that it will be freely draining, there will be a reasonable frictional strength, and there will not be problems owing to large creep deformations, as might be the case if a clay were used.

For the Vidal type of reinforced earth wall the distribution of tensile stresses along reinforcements has been found to be about as shown in Figure 20. The locus of maximum tensile stresses as a function of depth is also shown. This geometry can be used with soil property and reinforcement data and suitable factors of safety to select the length, cross-sectional area, and vertical and horizontal spacings of reinforcements. A similar methodology can be used for other reinforcing systems.

A generally accepted method for the seismic design of earth walls has not yet appeared. High-magnitude earthquakes were considered in the design of the walls at Valdez (Figure 9) and for other walls in areas of known seismicity. The design methodology led to a somewhat increased length of reinforcements and a higher density of reinforcements in the upper part of the wall. To the best of our knowledge, there have been no failures or significant distress of earth walls due to earthquakes.

Reinforcement durability is probably the area of greatest concern at present. The rate of corrosion of metal reinforcements depends on many factors, most of which cannot be controlled over the long term in the ground, e.g., local chemical concentrations and stray electrical currents. Galvanized steel has been extensively used. The by-products of zinc corrosion cover the base metal and tend to seal affected areas. A corrosion loss over the design life of the wall is taken into account in the specification of metal reinforcement cross sections. Epoxy-coated steel reinforcements now being developed offer the potential for high durability.

Nonmetallic reinforcing materials such as geotextiles, fiberglass, plastics, and composites, while not susceptible to corrosion, may undergo other chemical and/or biological forms of deterioration. Unfortunately, many of the materials are new, and the effects of long-term burial and exposure to the elements are not known. Hence, durability emerges as an area of major concern, and further studies are likely.

CONSTRUCTION

Generally the construction of earth walls is relatively simple, rapid, and straightforward. Equipment for placing and spreading the backfill soil and a roller for compacting it are required. Reinforcements can usually be carried and placed by hand, as shown in Figure 21. If precast facing panels are used, a small crane is required for handling them (Figure 22). Other facing types, e.g., geotextiles, do not require equipment for installation.

A significant advantage of reinforced earth walls over reinforced concrete walls is that no formwork is required. On the other hand, construction of a reinforced earth wall requires considerably greater space behind the wall face, as the reinforcement lengths are usually at least 0.7 times the wall height, whereas the base width of a concrete wall is only about 0.3 times the wall height.

COSTS

The costs for reinforced earth walls today are less, in constant dollars, than they were 8 to 10 years ago. This is because the technology is maturing and because there is competition among earth reinforcement systems. In spite of the widely different technologies, a parallel is evident with the electronics industry.

The total cost in any case will be composed of the costs of materials, erection, backfill soil (if the onsite soil is unsuitable), and any special aesthetic treatment of the facing that may be required. A reasonable figure for the cost of materials and erection for walls in the height range of 10 to 15 ft is about $15/ft^2$. For walls of 15 to 30 ft in height the cost increases to $17 or $18/ft^2$.

In urban areas reinforced concrete is likely to be more economical than reinforced earth for low walls up to 10 ft in height. The two types are competitive for wall heights of 10 to 30 ft, and the earth wall is less expensive for heights greater than 30 ft.

THE FUTURE

A number of reinforcing systems have been developed for construction of earth walls. Some are best suited for permanent high walls, some are best for low walls, some have their best application in remote areas, and others must withstand the rigors of high-traffic-volume urban areas.

As experience broadens and as uncertainties concerning the durability question disappear, it can be anticipated that earth reinforcement will see even greater use in surface transportation networks. The current competition between the manufacturers of different reinforcing materials and systems can lead to more efficient designs.

The rapid development and importance of the field has been recognized at the federal level. Work is in progress on National Cooperative Highway Research Project 24-2, Reinforcement of Earth Slopes and Embankments, administered by the National Research Council's Transportation Research Board. An international team of experts has been assembled to evaluate and compare the major systems and methods. The results of this study, due in late 1984, will be in the form of a report providing the user with the information required to make the analyses, designs, and estimates needed for the rational choice of an earth wall system.

BIBLIOGRAPHY

Bassett, R.H., and N.C. Last, "Reinforcing Earth Below Footings and Embankments," in *Proceedings of the Symposium on Reinforced Earth*, American Society of Civil Engineers, Pittsburgh, Pa., April 1978, pp. 202-231.

Binquet, J., and K.L. Lee, *Bearing Capacity of Strip Footings on Reinforced Earth Slabs*, G.I. 38983, Report to the National Science Foundation, Washington, D.C., May 1975.

Elias, V., and D.P. McKittrick, "Special Uses of Reinforced Earth in the United States," in *Proceedings of the International Conference on Soil Reinforcement: Reinforced Earth and Other Techniques*, Paris, Vol. 1, March 1979, pp. 255-259.

Goughnour, R.D., and J.A. Dimaggio, "Application of Reinforced Earth in Highways Throughout the United States," in *Proceedings of the International Conference on Soil Reinforcement: Reinforced Earth and Other Techniques*, Paris, Vol. 1, March 1979, pp. 301-313.

Maluche, E., "Reinforced Earth Used as Supporting Structures in Hydraulic Engineering," in *Proceedings of the International Conference on Soil Reinforcement: Reinforced Earth and Other Techniques*, Paris, Vol. 1, March 1979, pp. 335-339.

McKittrick, D.P., "Reinforced Earth: Application of Theory and Research to Practice," in *Proceedings of the Symposium on Reinforcing and Stabilizing Techniques* sponsored by the New South Wales Institute of Technology, October 1978. (Available from the Reinforced Earth Company, Rosslyn, Va.)

Mitchell, J.K., "Soil Improvement—State-of-the-Art Report," in *Proceedings of the Tenth International Conference on Soil Mechanics and Foundation Engineering*, Stockholm, Vol. 4, June 1981, pp. 509-565.

Schlosser, F., *Reinforced Earth*, ISSN 0337-1565, Ministère de l'Equipement, Laboratoire Central des Ponts et Chaussées, Paris, France, April 1976.

Vidal, H., "The Principle of Reinforced Earth," *Highw. Res. Rec., 282*:1-16, 1969.

FIGURE 1 Dry sand with maximum stable slope.

FIGURE 2 Moist sand can develop stable near-vertical slopes because of internal capillary water pressures.

FIGURE 3 Dry sand reinforced with strips of paper.

FIGURE 4 (Top) Ziggurat of Ur in Mesopotamia, circa 2500 B.C. Reinforcement is considered a major factor in its longevity. Photo Max Hirmer. (Bottom) Reconstructed view of the Ziggurat. Photo John Freeman. Reprinted from M. E. L. Mallowan, *Early Mesopotamia and Iran*, McGraw-Hill, New York, 1965, with permission.

FIGURE 5 Corduroy road for crossing swampy areas.

FIGURE 6 First highway use of modern reinforced earth wall—in France between Nice and the Italian border.

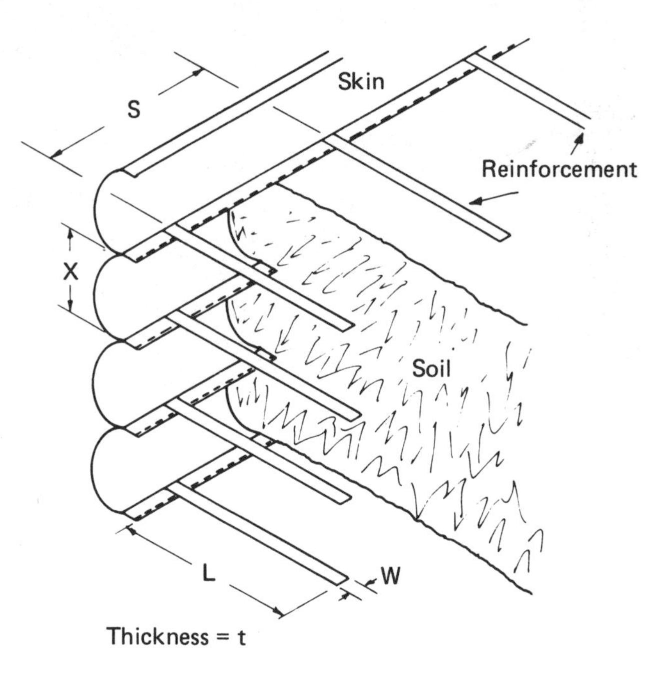

FIGURE 7 Concept of the Vidal reinforced earth wall.

FIGURE 8 First reinforced earth wall constructed in the United States, along Highway 39 in the San Gabriel Mountains of Southern California.

FIGURE 9 Reinforced earth walls used at Valdez, Alaska, in connection with the Alaska Oil Pipeline Terminal.

FIGURE 10 Special facing panels for reinforced earth wall along Interstate Highway 70 through Vail Pass, Colorado.

FIGURE 11 VSL Retained Earth system.

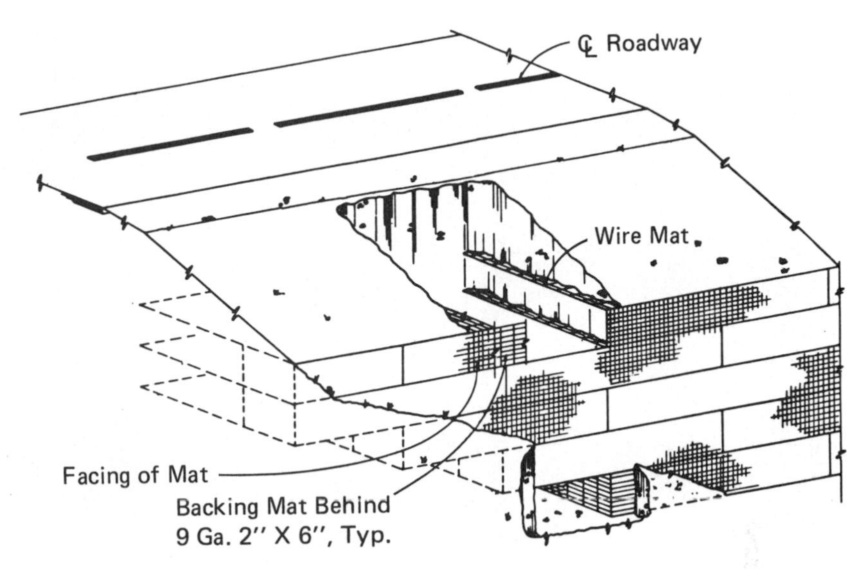

FIGURE 12 Schematic diagram of Hilfiker welded wire wall.

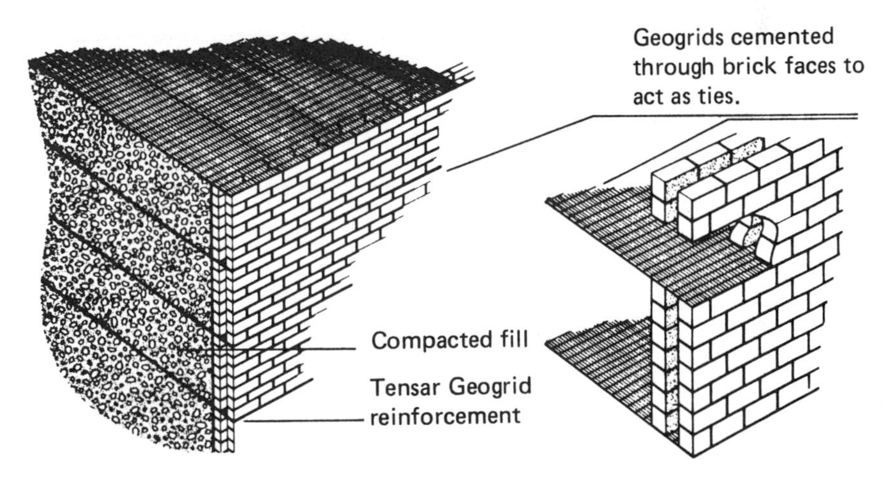

FIGURE 13 Tensar geogrids used to reinforce a brick retaining wall.

FIGURE 14 Fabric-reinforced earth wall.

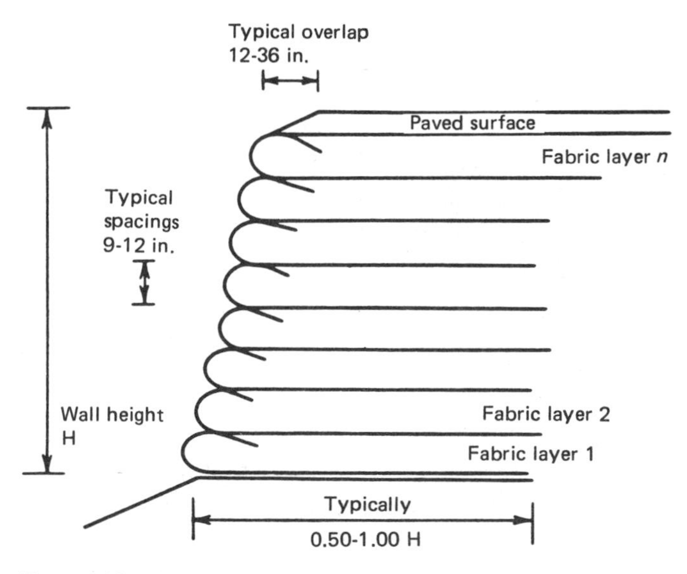

FIGURE 15 Schematic diagram of fabric-reinforced earth wall.

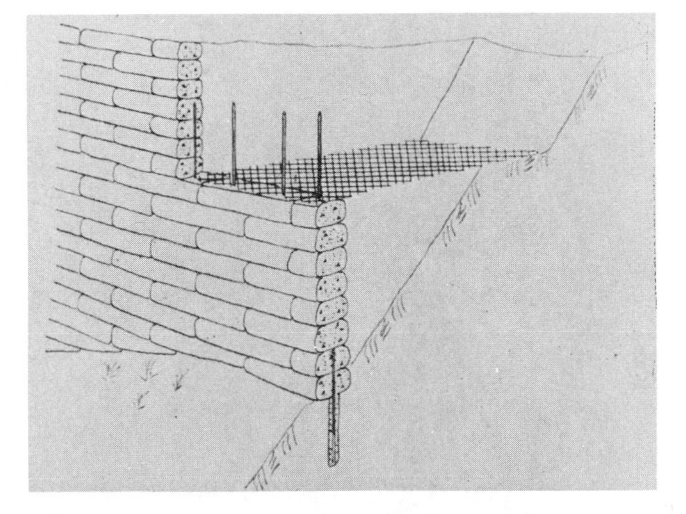

FIGURE 16 Schematic diagram of Kabil Stack-Sack Wall.

FIGURE 17 Soil nailing for improving cut slope stability.

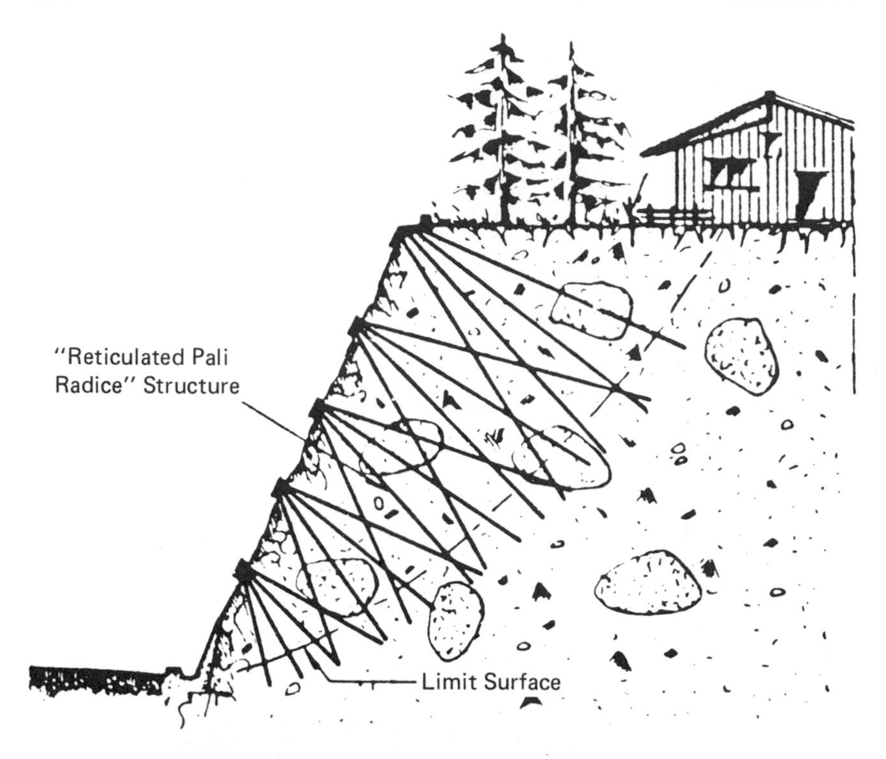

FIGURE 18 Root pile system for strengthening steep slope.

FIGURE 19 Reinforced earth bridge abutment.

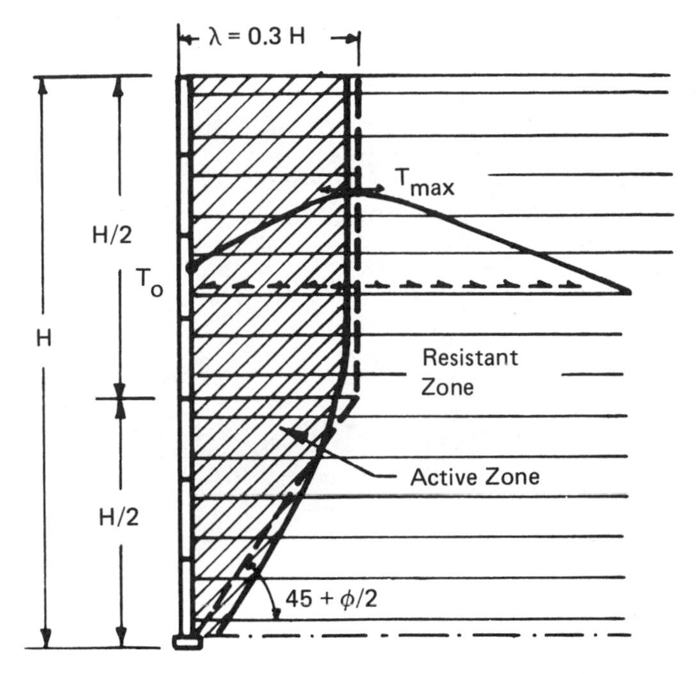

FIGURE 20 Stress distribution along reinforcement and locus of maximum tensile stresses in a reinforced earth wall.

FIGURE 21 Reinforcement placement.

FIGURE 22 Facing-panel installation.

Technological Opportunities and the Railroad Industry

WILLIAM J. HARRIS, JR.

Had one been discussing railroad technology 100 years ago, the railroads could have been characterized as being at the cutting edge. It took invention and major industrial change for them to become the great enterprise that they were and are. Without revolutions in the steel industry, the invention of the air brake system, and universities' dedicated mechanical engineering department studies of steam tables and energy conversion systems, the railroads would have simply remained a curiosity. As it was, this confluence of development created a system that shaped the economy of the United States.

The railroads became so successful that a whole set of national legislative and regulatory institutions had to be established. Some of these institutions lasted longer and stayed more rigid than we would have liked, but they were a consequence of enormous technological success.

The railroad industry today functions within an environment that is completely different from that of the past—today we have an extensive, mature transportation system and network. A completely rational society that made all economic decisions based on benefit costs would have provided a rather different set of institutions from those we have, which are a mixed set of public and private institutions. Public policy does indeed affect these; it provides major support to some elements of the transportation system and less to others. But what is given can also be taken away.

And so, in considering technological problems in the railroad industry, one must be extremely mindful of the context in which this mode of transportation is functioning and of the challenges to that mode. This economic issue helps to define the nature of relevant technological activities. The issue arises from competition between modes in the United States as well as from competition between foreign sources of supply of commodities produced for export in the U.S. economy.

Recent legislation in support of 48-foot trailers that are 102 inches wide and authorization for use of two trailers pulled by a single tractor establish a basis for significant cost reduction in highway transportation. The creation in South Africa of a very viable coal-exporting industry, one pending in China, one in Australia, and one in Poland are having profound effects on how the U.S. transportation system develops. They force us in technological directions that otherwise might not be pursued. These examples of domestic and foreign competition set the stage for a discussion of selected technologies that affect the railroad industry.

Although the dynamic interaction of the train with the track structure seems similar to other dynamic problems, there are some differences. There are more degrees of freedom and more complexities in the railroad train than in almost any other system. Trains carrying 14,000 tons in 140 or more cars, each car with four axles and eight wheels, each with two flexible connectors (couplers) at the ends operating over track structure that may be jointed every 39 feet, provides enormous potential for unsatisfactory dynamic interaction. There are many nonlinear elements in the system.

As a matter of fact, only a decade ago the designers of track did not understand the inputs to track from cars, and the designers of cars did not understand the effects of track perturbation on suspension systems and other features of cars. The engineer operating the locomotive had only the seat of his pants for guidance as to the correct control strategy.

That situation has improved. There are now mathematical models that permit running a train in a variety of ways over a variety of track structures with a variety of car designs, which has led to a $3-billion annual investment in maintenance of track structures. That in turn has led to the emergence of a new generation of cars designed to be more gentle on the track. And that development has led to precise guidance for the operators of trains and effective training programs that make possible much more relevant train operations.

As an analytical base has been built under these empirical relationships, a serious time constant has been encountered. A Track-Train Dynamics program was started in 1972; it will result in new components and new cars by 1985 or 1986; those cars will remain in the fleet until 2026. Thus, there is a 54-year time span from the beginning of research to the obsolescence of products based on that research. Over such a long period, there will unquestionably be changes in the nation's economy, in its societal structure, and in its regulatory structure. There is at present no effective response to the need for approaches that will identify these prospective changes which can be reflected in designs of equipment and systems. In other competitive modes, equipment becomes obsolete in 7 to 10 years. Thus, the other modes can remain more flexible in response to changing requirements.

The National Academy of Engineering has an opportunity to recognize some of these issues. It needs to look at the time for introduction of new technology, its life span, and the dynamics of changes in the economy that can affect new investments.

The study of dynamics is essential to railroad technology, and materials science and engineering are also important. The materials described in other papers in this volume represent great technological advances, but they are not "cents per pound" units, which the railroad industry requires. Nonetheless, there are improved, low-cost materials that are not in widespread use in the industry.

Combining our broader insights into dynamics and our review of low-cost, higher-performance materials, we can see an opportunity to restructure trains. Much weight can be taken out of each car. The weight of locomotives can be reduced but tractive effort retained by using powered axles under loaded cars. Very substantial redesign of train operating systems such as braking and coupling will be necessary. It is thought that those changes can reduce the costs of track operations by about 40 percent. If that can be done, new opportunities can be created for export of bulk commodities such as coal and grain, and thus a new demand for transportation can be developed. These commodities must now compete with lower-cost commodities available from foreign producers. If transportation costs can be reduced, our markets for these products can be retained and expanded.

Computers have been absolutely essential to the improved efficiency of the railroad industry. At present about 2 million freight cars operate with about 26,000 locomotives over about 285,000 miles of track structure in the United States. Two decades ago the procedures for keeping track of cars were unsatisfactory. That is no longer the case. With flexible computer programs and computer-to-computer communications, efficiency can be enhanced by providing lists of arriving cars to the next destination. And the industry is now applying microprocessors in a variety of ways; for example, in locomotives they identify the need for maintenance and the opportunity for more efficient control.

The actions of the Organization of Petroleum Exporting Countries (OPEC) have been devastating to the economics of every transportation mode. The oil bill of the railroads went from $350 million a year to $3.5 billion a year within about two years.

The Association of American Railroads has joined with the Southwest Research Institute, with the builders of locomotives, and with the oil industry to study all of the ways energy is used in the railroad system. For example, in a guided ground system with the flanged wheel and a rail, the flange may touch the rail on occasion, resulting in wear and loss of energy. The use of roller bearings improves safety, but the seals that keep the lubricants in place, without maintenance, for 7 to 10 years, bear on the axle and use energy.

In the past, with low-cost fuels, it was not necessary to pay attention to aerodynamic drag in relatively low-speed systems such as freight railroads. Now it is. Using fairings, closing freight car doors, and changing the spacing between trailers can reduce drag.

The locomotive builders have been able to reduce energy requirements by about 10 percent through a large number of changes in engine components.

Lubrication of the track to reduce the friction between the flange and the side of the head of the rail can reduce energy requirements by another 10 percent.

During the next few years we will probably abandon number 2 diesel and move toward a less expensive but still effective fuel. This can reduce costs by 10 percent. The cumulative effect of these changes will be a reduction of cost up to 25 percent.

Attention has been given to the interaction of these technologies with the labor force. The labor force in the railroad industry was 1.4 million in 1950; three years ago it was about 700,000; today it is under 400,000. Most of that reduction has occurred because of improved productivity. But the new technologies will provide further opportunities for improved production.

There are areas that need very real technical advances. Nondestructive inspection (NDI) of rail and wheels falls far short of what is technically required. As lighter and more highly stressed components are used, advanced structural design concepts will be employed that demand compatible NDI techniques.

Alternative energy conversion processes could make it feasible to use coal instead of liquid fuels, but none is immediately available to replace the diesel locomotive.

There are broad policy issues in transportation that need further examination. It is hoped that the National Research Council's Transportation Research Board will be able to clarify the demands that the emerging domestic economy will make on transportation. Until that guidance is available, public and private investment decisions may lead to extensive overcapacity and inefficient use of capital. The competitive technologies have to be brought into better focus than they are, not to create some broad national plan, but to provide a better basis for technical, engineering, and investment judgments.

The railroad industry, once extremely successful and now again extremely competitive with other modes, is working very hard to remain a dynamic force in the future of the economy. New technologies are contributing to that goal.

Biographical Sketches

CHARLES J. ARNTZEN, Director of the Michigan State University (MSU) Department of Energy Plant Research Laboratory, received bachelor's and master's degrees from the University of Minnesota and a Ph.D. degree from Purdue University. He served as a National Science Foundation postdoctoral fellow at the C. F. Kettering Research Laboratory. He was appointed to the Departments of Botany and Agronomy at the University of Illinois in Urbana and was Director of the campus-wide Cell Biology Program for two years. He was a research scientist with the U.S. Department of Agriculture (USDA). In 1980 he assumed his present position, and he also has a faculty appointment in the Biochemistry Department at MSU. Dr. Arntzen has received the Charles Albert Shull Award from the American Society of Plant Physiologists and the Superior Service Award from USDA; he is a member of the National Academy of Sciences.

J. PAUL BURNETT is Director of Molecular and Cell Biology Research at Eli Lilly and Company. He holds a Ph.D. in biochemistry from Indiana University and was a postdoctoral fellow at Harvard University Medical School. His responsibilities at Lilly Research Laboratories include the application of genetic engineering technology to pharmaceuticals and agricultural products. He has devoted much time to research on viruses that can transform animal cells, and his current interest is in recombinant DNA.

JAMES G. COLLIN is currently a graduate student working toward a Doctor of Engineering degree at the University of California, Berkeley. He received a B.S.C.E. from Union College in Schenectady, New York, and an M.S. from the George Washington University. Prior to his return to school for his doctorate, Mr. Collin worked for seven years as a foundation engineer for the George Hyman Construction Company. While employed, Mr. Collin held positions as both a design engineer and foundation superintendent and worked on numerous large-scale excavation projects.

CHARLES L. COONEY is Professor of Chemical and Biochemical Engineering at the Massachusetts Institute of Technology (MIT). He received his B.S. degree in chemical engineering from the University of Pennsylvania and took S.M. and Ph.D. degrees in biochemical engineering from MIT. He has served as a full-time consultant in fermentation technology for E. R. Squibb and Sons. He is author of more than 100 papers, coauthor or editor of two books, and acts as consultant to pharmaceutical, chemical, and biotechnological companies, as well as to various government and international organizations. Dr. Cooney serves as chairman of two chemical societies and was the recent recipient of the Food, Pharmaceutical, and Bioengineering Award from the American Institute of Chemical Engineers. His current research interests focus on problems of fermentation and enzyme technology and include computer control of biological processes, fuels, and chemicals production by fermentation, enzyme production, and scale-up of biotechnology processes.

MICHAEL L. DERTOUZOS is Professor of Electrical Engineering and Computer Science and Director of the Massachusetts Institute of Technology Laboratory for Computer Science.

WILLIAM J. HARRIS, JR., is Vice-President, Research and Test Department, Association of American Railroads (AAR). He received his B.S.Ch.E. and M.S.E. degrees from Purdue University and his Sc.D. in metallurgy from the Massachusetts Institute of Technology. Prior to joining the AAR, Dr. Harris was affiliated with Battelle and earlier with the National Academy of Sciences. He served as Secretary and Staff Director for the Presidential Task Force on Highway Safety in 1969. He has held offices in numerous professional organizations, has served on many councils, commissions, and advisory panels. He is a member of the National Academy of Engineering.

BERNARD H. KEAR is Scientific Advisor in Materials Science at Exxon's Corporate Research Center. He holds B.S. and Sc.D. degrees in metallurgy from the University of Birmingham, England. At the Franklin Institute in Philadelphia he studied the effects of long-range ordering on the plastic properties of crystals. At the Pratt & Whitney Division of United Technologies Corporation he investigated the interrelationships between structure and properties and processing in superalloys and participated in the development of single-crystal turbine blade technology. Dr. Kear has published approximately 120 technical papers, edited 6 books in the field of metallurgy, and has been awarded 22 patents in design and processing of superalloys. He was awarded the Mathewson Gold Medal of The Metallurgical Society of AIME and the Howe Medal of the American Society for Metals (ASM). He is a member of the National Academy of Engineering where he serves as Peer Committee chairman, a fellow of ASM, and is active in other professional organizations. He was the John Dorn Memorial Lecturer and is the 1983 Henry Krumb Memorial Lecturer. He is currently involved in research on rapid solidification and chemical vapor deposition.

JAMES D. MEINDL, Professor of Electrical Engineering at Stanford University, is Director of Stanford's Center for Integrated Systems. He holds a Ph.D. in electrical engineering from Carnegie-Mellon University. Following military duty with the U.S. Army Electronics Command at Fort Monmouth, New Jersey, he became Director of the Integrated Electronics Division there and was concurrently a lecturer on solid-state electronics at Monmouth College. He has received numerous awards for outstanding papers from Institute of Electrical and Electronics Engineers (IEEE) International Solid-State Circuits Conferences. In 1980 he was the recipient of the IEEE Electron Devices Society's J. J. Ebers Award for outstanding contributions to the field of electron devices. He is a member of the National Academy of Engineering and of other professional and honorary organizations. Dr. Meindl is the author of a book on micropower circuits and more than 250 technical papers. His current research interests focus on very large scale integration and integrated circuit applications in medical electronics.

JAMES K. MITCHELL is Professor and Chairman of the Department of Civil Engineering at the University of California, Berkeley. He holds a bachelor of civil engineering degree from Rensselaer Polytechnic Institute and master of science and doctor of science degrees from the Massachusetts Institute of Technology. He is the author of more than

170 published papers and a graduate-level text and reference, *Fundamentals of Soil Behavior*. Dr. Mitchell is a member of the National Academy of Engineering and a fellow of the American Society of Civil Engineers. He has received the Norman Medal, the Thomas A. Middlebrooks Award (three times), and the Walter L. Huber Research Prize of the American Society of Civil Engineers. He has also received the Western Electric Fund Award of the American Society for Engineering Education and the Medal for Exceptional Scientific Achievement from the National Aeronautics and Space Administration. His research activities have included studies of soil mechanics, various methods of soil improvement and reinforcement, physicochemical phenomena in soils, the stress-strain time behavior of soils, lunar soil mechanics, and in situ measurement of soil properties.

ROGER S. PORTER is with the Graduate Research Center of the University of Massachusetts at Amherst. He has B.S. and Ph.D. degrees in chemistry from the University of California at Los Angeles and the University of Washington at Seattle, respectively. He was with the Chevron Research Company, and at the University of Massachusetts he has held the following positions: Program Chairman of Polymer Science and Engineering, Codirector of the National Science Foundation Materials Research Laboratory, and Special Research Assistant to the University President. He received an award for polyolefin research, the International Award in Education, and the International Award in Plastics Science and Engineering from the Society of Plastics Engineers, and for Organic Coatings and Plastics from the American Chemical Society. He also received the Meritorious Service Award from the Plastics Institute of America and the Mettler Award from the North American Thermal Analysis Society. Dr. Porter serves on the Board of Trustees, Gordon Research Conferences, and is currently Chairman.

HERBERT SCHORR is Research Vice-President for Systems at the International Business Machines Corporation and is responsible for computer science activities at sites in the United States, in Zürich, Switzerland, and at the Japan Science Institute. His undergraduate education was at City University of New York, and he holds a Ph.D. in electrical engineering from Princeton University. He has served as an instructor of electrical engineering at Princeton and as an assistant professor at Columbia University. He was a National Science Foundation postdoctoral fellow at Cambridge University in England. Dr. Schorr joined IBM as a research staff member and has progressed to his present position. He is a member of the Institute of Electrical and Electronics Engineers and the Association for Computing Machinery.

JAN P. SKALNY is Associate Director of Martin Marietta Laboratories, responsible for research in advanced ceramics technology and occupational health and analytical chemical services. He has published more than 80 technical papers, organized scientific meetings, and consulted/lectured around the world. He is a Fellow of the American Ceramic Society, a member of editorial boards of several scientific journals, and has participated in National Research Council studies on cement R&D and solidification of radioactive wastes.

JOHN E. STEINER has recently retired as Vice-President for Corporate Product Development of the Boeing Company. He is currently serving as a consultant to both the industry and the federal government on significant matters involving national aeronautics policy. He received his undergraduate degree from the University of Washington and his master of science degree in aeronautical engineering from the Massachusetts Institute of Technology. During his 40 years with Boeing, Mr. Steiner was actively involved in the development of virtually all Boeing airplanes. He was selected as the sole industry participant in the White House aeronautics policy study of 1982. He is a member of the National Academy of Engineering and is currently a panel member of the National Research Council's study on The Competitive Status of the U.S. Civil Aviation Manufacturing Industry. He holds the rank of fellow and honorary fellow in numerous foreign aeronautical societies, and serves on many boards and commissions. He has twice been named "Man of the Year" by *Aviation Week* and has received the Elmer A. Sperry Award, Australia's Sir Charles E. Kingsford-Smith Memorial Medal, the Thulin Medal of Sweden, and the University of Washington's highest alumni honor, the Summa Laude Dignatus designation.

ALBERT R. C. WESTWOOD is Corporate Director for Research and Development, Martin Marietta Corporation. He received his B.Sc., Ph.D., and D.Sc. degrees from the University of Birmingham, England. He joined Martin Marietta Laboratories (then RIAS) in 1958, becoming its Director in 1974 and assuming his present position in January 1984. He has published well over 100 technical papers, mostly concerned with environment-sensitive mechanical behavior and, lately, R&D management, and presented numerous keynote and invited lectures around the world. His scientific contributions have been recognized by a variety of awards and fellowships, including the Beilby Gold Medal (1970) and election to the National Academy of Engineering (NAE) (1980). His current professional responsibilities include the National Materials Advisory Board, the Board of Directors of the Industrial Research Institute, the Board on Science and Technology for International Development, the Advisory Councils to the School of Arts and Sciences at the Johns Hopkins University and School of Engineering at Maryland, and Alternate to the Foreign Secretary of the National Academy of Engineering.